Mathematics for Industry

Volume 24

Aims & Scope

The meaning of "Mathematics for Industry" (sometimes abbreviated as MI or MfI) is different from that of "Mathematics in Industry" (or of "Industrial Mathematics"). The latter is restrictive: it tends to be identified with the actual mathematics that specifically arises in the daily management and operation of manufacturing. The former, however, denotes a new research field in mathematics that may serve as a foundation for creating future technologies. This concept was born from the integration and reorganization of pure and applied mathematics in the present day into a fluid and versatile form capable of stimulating awareness of the importance of mathematics in industry, as well as responding to the needs of industrial technologies. The history of this integration and reorganization indicates that this basic idea will someday find increasing utility. Mathematics can be a key technology in modern society.

The series aims to promote this trend by (1) providing comprehensive content on applications of mathematics, especially to industry technologies via various types of scientific research, (2) introducing basic, useful, necessary and crucial knowledge for several applications through concrete subjects, and (3) introducing new research results and developments for applications of mathematics in the real world. These points may provide the basis for opening a new mathematics-oriented technological world and even new research fields of mathematics.

More information about this series at http://www.springer.com/series/13254

Yoshinori Dobashi · Hiroyuki Ochiai
Editors

Mathematical Progress in Expressive Image Synthesis III

Selected and Extended Results from the Symposium MEIS2015

Springer

Editors
Yoshinori Dobashi
Hokkaido University
Sapporo, Hokkaido
Japan

Hiroyuki Ochiai
Kyushu University
Fukuoka
Japan

ISSN 2198-350X ISSN 2198-3518 (electronic)
Mathematics for Industry
ISBN 978-981-10-1075-0 ISBN 978-981-10-1076-7 (eBook)
DOI 10.1007/978-981-10-1076-7

Library of Congress Control Number: 2014939620

Printed on acid-free paper

This Springer imprint is published by Springer Nature
The registered company is Springer Science+Business Media Singapore Pte Ltd.

Preface

The international symposium "Mathematical Progress in Expressive Image Synthesis" (MEIS), provides a unique venue where mathematicians, computer graphics (CG) researchers, and those who work in industry gather to share and discuss shared contemporary issues and future collaborative projects. The international symposium MEIS2015 aimed to furnish a venue where various issues in CG application fields could be discussed by mathematicians, CG researchers, and practitioners. Through the previous conferences, MEIS2013 and MEIS2014, mathematicians as well as CG researchers have recognized that CG is a specific and practical activity derived from mathematical theories. Issues found in CG broaden the field of mathematics (and vice versa), and CG visualizes mathematical theories in an aesthetic way.

This volume presents the papers selected from the MEIS2015 proceedings, originally issued as MI Lecture Notes Vol. 64, Kyushu University, 2015. Several invited talks attracted and inspired prospective attendees who work in academia or industries having strong interests in digital media creations, scientific visualization, and visual engineering. This year, we aimed to provoke interdisciplinary research projects through the peer-reviewed paper/poster presentations at the symposium. The topics included geometry, curves and surfaces, fabrication, fluid, interpolation, illusion, texture, visualization, and rendering. We hope readers will find themselves deeply inspired through the harmony of mathematics and graphics research and the industrial work displayed in this volume.

February 2016

Yoshinori Dobashi
Hiroyuki Ochiai

MEIS2015 Committee

Co-Chairs

Hiroyuki Ochiai, Kyushu University
Yoshinori Dobashi, Hokkaido University

International Program Committee

Ken Anjyo, OLM Digital
Jean-Marie Aubry, Weta Digital
Bernd Bickel, Disney Research
Yasuhide Fukumoto, Kyushu University
Sunil Hadap, Adobe Research
Kei Iwasaki, Wakayama University
Shizuo Kaji, Yamaguchi University
Kenji Kajiwara, Kyushu University
Miyuki Koiso, Kyushu University
J.P. Lewis, Victoria University of Wellington/Weta Digital
Yoshihiro Mizoguchi, Kyushu University
Yoshiyuki Ninomiya, Kyushu University
Yutaka Ohtake, The University of Tokyo
Makoto Okabe, University of Electro-Communications
Daisuke Tagami, Kyushu University

External Reviewer

Alexandre Derouet-Jourdan, OLM Digital

Acknowledgments

The MEIS2015 Organizing Committee is very much grateful to the Institute of Mathematics for Industry (IMI), Kyushu University for sponsoring the Symposium MEIS2015. We would like to thank the Japan Science and Technology Agency (JST), Mathematics Program: Alliance for Breakthrough between Mathematics and Sciences (ABMS) on our five-year project "Mathematics for Computer Graphics" for continuous support. We also extend our thanks to Kazuko Ito for her hard work on the conference arrangement, Seiko Sasaguri and Nagatoshi Sasano on the production of the proceedings, and Ayumi Kimura for both. Last but not least, we appreciate the hard work of the international program committee and the external reviewers in the tight schedule.

Contents

Geometry and Mechanics of Fibers: Some Numerical Models

Florence Bertails-Descoubes

Abstract In this talk I will give an overview of our work on the simulation of fibers and entangled materials, such as hair, with a specific interest for virtual prototyping and computer graphics applications. I will first introduce a family of high-order, reduced models for discretizing Kirchhoff's equations for thin elastic rods in a both faithful and robust way. Such models are particularly well-suited for simulating inextensible fibers subject to bending and twisting, and featuring an arbitrary curly resting geometry. Then I will show how such models can be coupled to frictional contact using the nonsmooth contact dynamics framework, and I will present a hybrid iterative solver suitable for robustly handling thousands packed fibers at reasonable frame rates. Finally, I will give some insights into the inverse modeling of fibers, consisting in taking an arbitrary curve geometry as input and inferring corresponding geometric and physical parameters of the simulator such that the input geometry corresponds to a stable configuration at equilibrium.

Keywords Physics-based simulation · Thin elastic rod · Frictional contact · Hair simulation · Inverse physics-based design

1 Introduction

Deformable slender structures such as hair fibers, cloth, ribbons, tree branches or leaves, are ubiquitous around us. They often feature an intricate natural shape, ranging from straight to curly, and are characterized by a complex motion involving strongly nonlinear deformations, such as buckling. These complex shapes and motions greatly contribute to the visual richness of the real world. When multiple such structures are

---

(Joint work with Romain Casati, Gilles Daviet, and Alexandre Derouet-Jourdan).

F. Bertails-Descoubes (✉)
Inria - Laboratoire Jean Kuntzmann (CNRS - Grenoble University) Inria Rhône-Alpes, 655 Avenue de l'Europe, 38334 Saint-Ismier Cedex, France
e-mail: Florence.Descoubes@inria.fr

Y. Dobashi and H. Ochiai (eds.), *Mathematical Progress in Expressive Image Synthesis III*, Mathematics for Industry 24,
DOI 10.1007/978-981-10-1076-7_1

coupled together with contact and friction, the range of emerging phenomena is even more exacerbated, giving rise to stick-slip dynamical instabilities, entangling, or spontaneous collective behavior. Human hair, which is typically composed of 150,000 thin fibers, beautifully depicts such complex mechanical behavior when fluttering in the wind.

As the essence of Computer Graphics is to represent the visual appearance of the real world with the highest fidelity, it is important for practitioners to be able to capture all the relevant details of a given scene. In the case of a dynamical scene involving passive objects such as hair, cloth, or natural phenomena, *physics-based simulation* has proven over the years to be a method of choice for capturing resulting visual effects. Unlike phenomenological methods which develop descriptive models for reproducing a given emerging phenomenon, physics-based methods provide generative models whose goal is to explain the physical causes of the phenomenon. From a set of initial conditions as well as a few physical parameters (e.g., the mass, the stiffness, the natural shape), a physics-based simulator may generate not just a single effect, but a wide range of emerging phenomena revealing the whole complexity of the underlying physics.

When designing a physics-based simulator for computer graphics, one has inevitably in mind the four following criteria:

1. **Realism**: Ingredients which are necessary to capture relevant visual effects should be identified, and translated numerically with as few quality loss as possible.
2. **Robustness**: The simulator should converge properly for a subsequent range of parameters.
3. **Efficiency**: The simulator should be fast enough for allowing complex scenes to be simulated in reasonable timings (in our case, a few days of computation for a given scene is considered as an upper-bound).
4. **Control**: The simulator should provide the user with some handles to control the shape and motion of the object in an intuitive way.

Over the ten past years, we have been striving to develop some numerical models satisfying all four criteria at the same time. Our work has focused on the simulation of slender structures prone to contact and dry friction, and especially on the dynamics of thin fiber assemblies, with some direct applications in hair simulation and inverse physics-based design.

2 High-Order Reduced Models for Simulating Dynamic Fibers

The first part of the talk will be devoted to the presentation of the numerical models that we have been developing for simulating the dynamics of flexible fibers, the so-called *super-helix* and *super-clothoid* models [1, 2, 4].

I will first introduce the mechanical equations for inextensible thin elastic rods, namely the *Kirchoff* equations, which take the form of second order partial differential equations subject to boundary conditions. Noting that curvatures and twist play a major role both in the geometric and dynamic description of this model, we have come up with a spatial rod discretization based on elements that are polynomial in such quantities.

Our first scheme relied on piecewise constant curvatures [1], and was then extended to piecewise linear curvatures [2, 4]. One major advantage of such curvature-based formulations is that the kinematics of the discretized rod remains, by construction, perfectly inextensible. Such intrinsic inextensibility thus removes the burden of adding subsequent (stiff) inextensibility contraints when solving the dynamics. The price to pay, however, is that the geometry of the rod is not readily available but has to be computed iteratively from the curvatures.

In the piecewise constant case [1], each element turns out to be a perfect circular helix (hence the super-helix name for the model), leading to a cheap and exact evaluation of all the terms of the discrete dynamic equations. For higher orders however, one unfortunately loses such a closed-form formula and a both accurate and efficient spatial integration scheme has to be designed.

In the piecewise linear case (where each element takes the form of a 3D clothoid), we were able to build an accurate integration scheme which proved to be orders of magnitude faster compared to classical integration methods [4]. The key of our approach was to leverage the fact that the solution takes the form of a power series expansion, while avoiding the pitfall of catastrophic cancellation through an adaptive integration strategy. With this tool in hand, we were able to demonstrate that the super-clothoid model could capture intricate shapes both robustly and efficiently, with better spatial accuracy and geometric fairness compared to state-of-the-art methods (see Fig. 1).

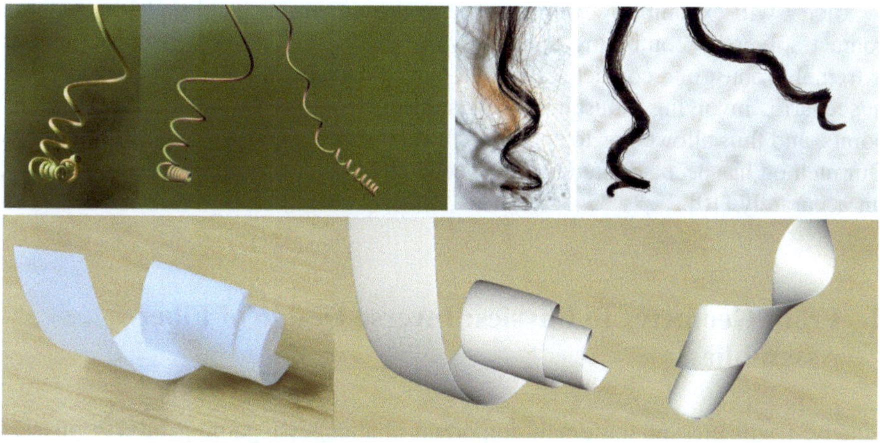

Fig. 1 Many physical strands exhibit a smooth curled geometry with linear-like curvature profile, which is captured and deformed accurately thanks to our super-clothoid model [4]. From *left* to *right* and *top* to *bottom*, three examples of real strands whose shapes are synthesized and virtually deformed in real-time using a very low number of 3D clothoidal elements: a vine tendril (4 elements), a hair ringlet (2 elements), and a curled paper ribbon (1 single element). *Left* photograph courtesy of Jon Sullivan, pdphoto.org.

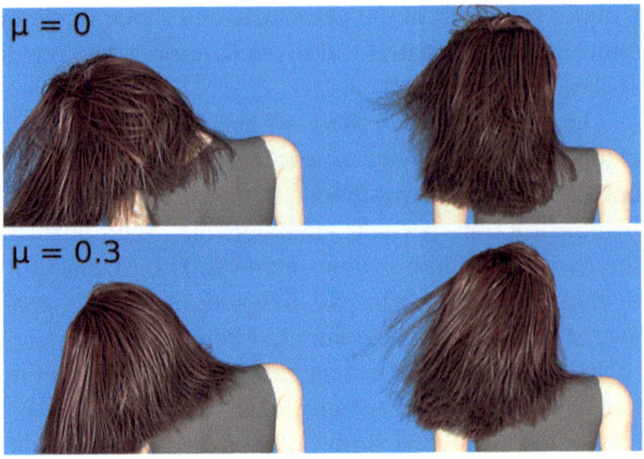

Fig. 2 Simulation of a fast head movement without (*top*) and with (*bottom*) Coulomb friction, using our robust solver from [5]. In the latter case, hair remains much more coherent and depicts typical stick-slip effects

3 Robust Frictional Contact Model for Fiber Assemblies

The second part of the talk will be focused on the dynamic simulation of fiber assemblies, where individual fibers are coupled to each other through contact and friction.

I will first illustrate why capturing threshold effects in friction is key to realism (see Fig. 2), before introducing the nonsmooth *Signorini-Coulomb* friction model and its various formulations. We shall see what the numerical counter-parts are for each formulation, and how each of them performs in terms of efficiency, robustness, and scalability [3].

Then I will explain how we managed to design a robust and scalable frictional contact solver by combining an iterative Gauss-Seidel strategy together with an extremely robust one-contact solver [5]. Our global solver proved to converge well in scenarios involving thousands fibers subject to tens thousands frictional contact points, and thus allowed us to enhance considerably the realism of hair simulations. Our method has been adopted by the special effects industry for simulating hair and fur accurately [10].

4 From Geometry to Physics: Inverse Design of Fiber Assemblies

Finally, I will present some new important challenges regarding inverse physics-based design. While current simulators may succeed in reaching a good level of realism, they remain difficult to control in order to achieve a precise artistic goal. More precisely, to generate some desirable shapes and motions, one should be able to feed a simulator with the "right" parameters. Finding such parameters remains a difficult task, which is often performed through a tedious trial and error process.

To make this task fully automatic, we have started looking at *inverse* solutions in the case where a static shape is provided as input: the inverse model should be able to interpret this shape automatically as a stable equilibrium of the simulator under gravity and other external forces, such as contact and friction.

In the case of an isolated fiber, we have shown that inverting any of our super-model [1, 4] boils down to two decoupled problems, both of them being easy to solve [6, 7]: First, an equilibrium condition which appears to be linear in the natural shape of the fiber, thanks to the curvature-based parameterization of our fiber models; Second, a sufficient stability condition that can be simply set by fixing a lower-bound for the ratio of stiffness over mass. Actually, the only remaining difficulty is to solve a merely geometric fitting problem—converting a curve as a piecewise helix or clothoid. In the case of helical fitting, we have already brought some efficient and robust solutions to this geometric approximation problem [6, 9].

In the presence of contact and friction, Coulomb sticking constraints have to be considered, which makes the overall inverse problem nonsmooth and ill-posed. We have shown that assuming known mass and stiffness, as well as a simplified inverse model, it is possible to recover a plausible natural shape as well as frictional contact forces at play [8]. This work allowed us, for the first time, to animate in a plausible way a few hair geometries stemming from recent hair captures (see Fig. 3).

Fig. 3 Real curly wig (*left*) captured from [11], inverted by our method in [8] and physically animated (*middle*) and trimmed (*right*)

5 Conclusion

Throughout this long-term work on the numerical modeling of fibers and frictional contact, we have learnt that systematically concentrating the efforts on the upstream modeling and formulation of problems often pays off: even for very complex problems, the resulting numerics may be greatly simplified, and thus solved more easily and robustly. Keeping in mind this key lesson, we are starting to investigate the case of 2D slender structures (namely plates and shells), for which many exciting challenges remain open.

References

1. F. Bertails, B. Audoly, M.-P. Cani, B. Querleux, F. Leroy, J.-L. Lévêque, Super-helices for predicting the dynamics of natural hair. ACM Trans. Gr. (Proc. ACM SIGGRAPH'06) **25**, 1180–1187 (2006)
2. F. Bertails-Descoubes, Super-clothoids. Comput. Gr. Forum (Proc. Eurographics'12) **31**(2pt2), 509–518 (2012)
3. F. Bertails-Descoubes, F. Cadoux, G. Daviet, V. Acary, A nonsmooth Newton solver for capturing exact Coulomb friction in fiber assemblies. ACM Trans. Gr. **30**, 6:1–6:14 (2011)
4. R. Casati, F. Bertails-Descoubes, Super space clothoids. ACM Trans. Gr. (Proc. ACM SIGGRAPH'13) **32**(4), 48:1–48:12 (2013)
5. G. Daviet, F. Bertails-Descoubes, L. Boissieux, A hybrid iterative solver for robustly capturing Coulomb friction in hair dynamics. ACM Trans. Gr. (Proc. ACM SIGGRAPH Asia'11) 30, 139:1–139:12 (2011)
6. A. Derouet-Jourdan, F. Bertails-Descoubes, J. Thollot, Stable inverse dynamic curves. ACM Trans. Gr. (Proc. ACM SIGGRAPH Asia'10) 29, 137:1–137:10 (2010)
7. A. Derouet-Jourdan, F. Bertails-Descoubes, J. Thollot, 3D Inverse dynamic modeling of strands, ed. by D. Wexler, ACM SIGGRAPH 2011 Posters, page Article No. 55, Vancouver, Canada, August 2011. ACM SIGGRAPH, ACM. Poster (2011)
8. A. Derouet-Jourdan, F. Bertails-Descoubes, G. Daviet, J. Thollot, Inverse dynamic hair modeling with frictional contact. ACM Trans. Gr. **32**(6), 159:1–159:10 (2013)
9. A. Derouet-Jourdan, F. Bertails-Descoubes, J. Thollot, Floating tangents for approximating spatial curves with G^1 piecewise helices. Comput. Aided Geom. Des. **30**(5) 490–520 (2013)
10. D. Kaufman, R. Tamstorf, B. Smith, J.-M. Aubry, E. Grinspun, Adaptive nonlinearity for collisions in complex rod assemblies. ACM Trans. Gr. (SIGGRAPH 2014) 2014
11. L. Luo, H. Li, S. Rusinkiewicz, Structure-aware hair capture. ACM Trans. Gr. (Proc. ACM SIGGRAPH'13) **32**(4), 76:1–76:12 (2013)

Tetrisation of Triangular Meshes and Its Application in Shape Blending

Shizuo Kaji

Abstract The As-Rigid-As-Possible (ARAP) shape deformation framework is a versatile technique for morphing, surface modelling, and mesh editing. We discuss an improvement of the ARAP framework in a few aspects: 1. Given a triangular mesh in 3D space, we introduce a method to associate a tetrahedral structure, which encodes the geometry of the original mesh. 2. We use a Lie algebra based method to interpolate local transformation, which provides better handling of rotation with large angle. 3. We propose a new error function to compile local transformations into a global piecewise linear map, which is rotation invariant and easy to minimise. We implemented a shape blender based on our algorithm and its MIT licensed source code is available online.

Keywords Shape blending · Tetrahedral mesh · As-rigid-as-possible deformation

1 Introduction

In Shape blending the seminal paper [1], they introduced a morphing algorithm called the As-Rigid-As-Possible (ARAP, for short) shape interpolation. Since then, the technique has been successfully applied to various shape deformation applications. In their original paper, tetrahedral volume meshes are used to produce interpolation of shapes. However, in most computer graphic systems it is common to represent shapes by surface meshes. To convert a surface mesh to a volume mesh is a non-trivial task (see, for example, [13]) and the resulting volume mesh tends to have many extra internal vertices, which makes applications inefficient. Instead of considering volume meshes, one can "fatten" surface meshes. A common practice is to associate a tetrahedral structure to a triangular surface mesh by adding the normal vector for every triangle (see, for example, [15]). Although this simple trick has been widely

S. Kaji (✉)
Yamaguchi University/JST CREST, 1677-1, Yoshida, Yamaguchi 753-8512, Japan
e-mail: skaji@yamaguchi-u.ac.jp

Y. Dobashi and H. Ochiai (eds.), *Mathematical Progress in Expressive Image Synthesis III*, Mathematics for Industry 24,
DOI 10.1007/978-981-10-1076-7_2

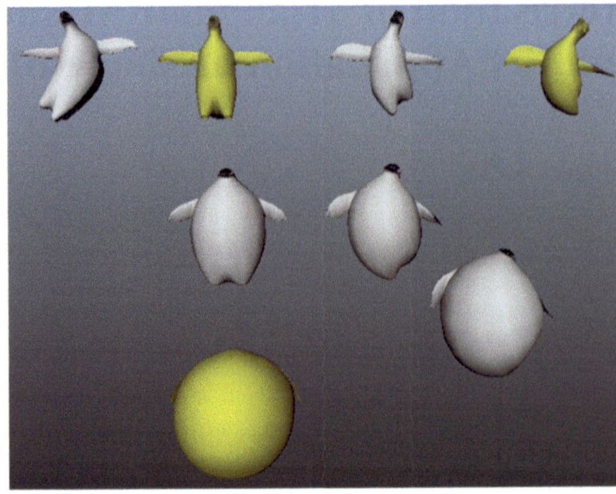

Fig. 1 Three shapes (*yellow*) are blended to produce variations (*white*). Not only interpolation but also extrapolation (with weights >1 or <0) is possible. The *top-left shape* is obtained by extrapolating the two *yellow shapes* in the *top row*

used, it does not capture important geometric features of the mesh. For example, the relation between adjacent triangles is neglected.

One of the main purposes of this paper is to introduce a new construction to associate a tetrahedral structure to a triangular mesh, which we call *tetrisation* (Sect. 4). Our method encodes inter-triangular properties such as the angle between adjacent triangles so that one can keep track of global geometry such as curvature while working locally on tetrahedra.

We also discuss an improvement of the ARAP (Sect. 3) in how to interpolate local transformations (Sect. 5) and how to stitch fragmented tetrahedra by a new error function (Sect. 6). We demonstrate our improvement by a shape blending application (Fig. 1). Given an arbitrary number of isomorphic surfaces, our algorithm produces inter/extrapolation of the shapes according to the weights given by the user. Roughly speaking, we define a "linear combination" of shapes

$$w_1 Q_1 + w_2 Q_2 + \cdots + w_m Q_m,$$

where $w_i \in \mathbb{R}$ are weights and Q_i are shapes. In particular, when the number of shapes is two, $w_1 Q_1 + (1 - w_1) Q_2$ for $0 \leq w_1 \leq 1$ gives a morphing between them. Note that our algorithm is highly non-linear although we described the procedure as taking the linear combination of shapes. We implemented the algorithm as the Autodesk Maya plugin. Its MIT licensed source code is available at [6].

2 Notation

We begin with listing some notation. We assume all transformations are represented by real matrices, acting on real column vectors by the multiplication from the left.

- SO(3): the group of 3D rotations. Its element is a 3×3 special orthogonal matrix.
- $\mathrm{Sym}^+(3)$: the set of 3D shears. Its element is a 3×3 positive definite symmetric matrix.
- GL(3): the group of (invertible) 3D linear transformations consisting of compositions of rotation, shear, and reflection. Its element is a 3×3 regular matrix.
- Aff(3): the group of (invertible) 3D affine transformations consisting of compositions of rotation, shear, reflection, and translation. Its element is a 4×4 regular homogeneous matrix.
- $\mathrm{GL}^+(3)$, $\mathrm{Aff}^+(3)$: the subgroups of the reflection free (positive determinant) elements in the corresponding groups.
- $\hat{A} \in \mathrm{GL}(3)$: the linear part ($3 \times 3$ upper-left corner) of $A \in \mathrm{Aff}(3)$.
- A^t: the transpose of a matrix A.
- $|A|_F^2 = \mathrm{tr}(A^t A)$: the squared Frobenius norm of a matrix A.
- $\#U$: the cardinality of a set U.

3 As-Rigid-As-Possible Deformation Framework

In this section, we recall the ARAP framework by describing an algorithm for shape blending. Note that although we discuss shape blending as the primary application, the framework and our improvement is not limited to it. Indeed, after being introduced in [1] initially as a morphing algorithm, the ARAP technique has been serving as one of the fundamental frameworks for various kinds of shape deformation applications (see, for example, [3, 7, 14–16]).

Our problem setting is as follows. We are given a rest shape V_0 and m its deformations V_j $(1 \le j \le m)$. That is, a vertex correspondence between V_0 and each of V_j $(1 \le j \le m)$ is assumed. We would like to compute the deformation $V(w_1, \dots, w_m)$ by blending the given shapes $\{V_j\}$ according to the user specified weights $\{w_j \in \mathbb{R} \mid 1 \le j \le m\}$. We insist that it interpolates the given shapes, i.e., $V(0, \dots, 0) = V_0$, and $V(w_1, \dots, w_m) = V_k$ when $w_j = \begin{cases} 1 & (j = k) \\ 0 & (j \neq k) \end{cases}$. Notice we allow negative weights and weights greater than one so that the system can not only interpolate but also extrapolate.

Remark 1 A basic shape blending is achieved by simply taking the linear combination of the coordinates of the vertices. This method is very fast and widely used to produce variations of shapes, in particular, facial expressions. However, since the geometry of shapes is disregarded, it does not always produce plausible outputs (Fig. 2). The ARAP based method which we will describe below takes geometry into account to obtain better results.

We assume that the rest shape is equipped with a non-degenerate *tetrahedral structure* $(V_0, \mathcal{T})$. We will discuss in Sect. 4 a method to associate one to a triangular mesh.

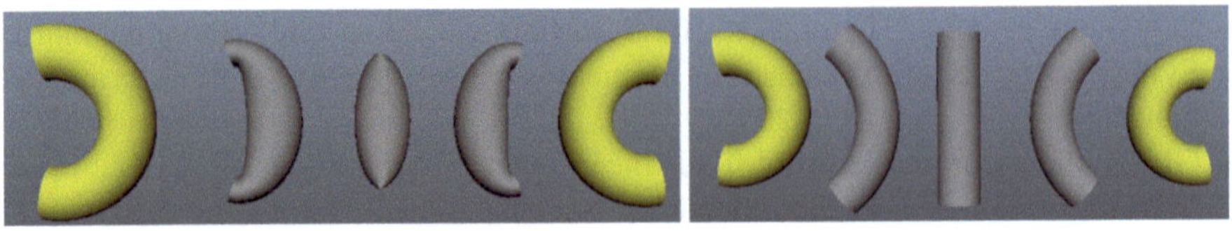

Fig. 2 Interpolation between *yellow shapes*. *Left* linear method. *Right* our method

Definition 1 A tetrahedral structure is a pair $(V, \mathscr{T})$, where the vertex set V consists of three dimensional vectors and the set of tetrahedra $\mathscr{T} = \{T_i \mid 1 \leq i \leq n\}$ consists of ordered tuples of four distinct vertices $T_i = (v_{i_1}, v_{i_2}, v_{i_3}, v_{i_4})$. Each vertex in V must be contained in at least one tetrahedron. A tetrahedral structure is said to be non-degenerate when the vertices of each tetrahedron are not co-planar.

We emphasise that a triangle can be shared by three or more tetrahedra, and for this reason, we use the terminology "tetrahedral structure" rather than tetrahedral mesh.

The information of a tetrahedral structure $(V, \mathscr{T})$ can be packed into a collection of 4×4-matrices:

$$\left\{ P_i := \begin{pmatrix} v_{i_1}(x) & v_{i_2}(x) & v_{i_3}(x) & v_{i_4}(x) \\ v_{i_1}(y) & v_{i_2}(y) & v_{i_3}(y) & v_{i_4}(y) \\ v_{i_1}(z) & v_{i_2}(z) & v_{i_3}(z) & v_{i_4}(z) \\ 1 & 1 & 1 & 1 \end{pmatrix} \middle| 1 \leq i \leq n \right\}, \tag{1}$$

where $(v_{i_j}(x), v_{i_j}(y), v_{i_j}(z))^t \in \mathbb{R}^3$ is the vector representing the position of the vertex $v_{i_j} \in V$.

We denote by $\{P_{0i} \mid 1 \leq i \leq n\}$ the matrices associated to the rest shape $(V_0, \mathscr{T})$. Since $(V_0, \mathscr{T})$ is assumed to be non-degenerate, all the P_{0i} are regular. For each deformation V_j, we use the same set of tetrahedra $\mathscr{T}$ to obtain $\{P_{ji} \mid 1 \leq i \leq n\}$. Note that P_{ji} need not be regular. We define a series of affine transformations

$$A_{ji} := P_{ji}P_{0i}^{-1} \ (1 \leq i \leq n) \tag{2}$$

which maps the vertices V_0 of the rest shape to the ones V_j in the deformed shape. Obviously, $A_{ji}v = A_{ji'}v$ when $v \in V_0$ is contained in two tetrahedra T_i and $T_{i'}$. Thus, $\{P_{ji} \mid 1 \leq i \leq n\}$ can be considered as a *piecewise linear map* defined on $(V_0, \mathscr{T})$ with $(V_j, \mathscr{T})$ as its image.

Now, we have m piecewise linear maps $\{P_{ji} \mid 1 \leq i \leq n\}_{j=1}^m$ and the problem is rephrased as to blend them according to the user specified weights w_j $(1 \leq j \leq m)$. We first consider locally and blend $\hat{P}_{ji}$ $(1 \leq j \leq m)$ for a single tetrahedron T_i to obtain $C_i \in \mathrm{GL}^+(3)$. Intuitively, C_i stipulates the local transformation for the tetrahedron T_i. We discuss a method to compute C_i in Sect. 5. The last step is to find a global piecewise linear map on $(V_0, \mathscr{T})$, whose image we take as the output. Since we cannot assume $C_i v$ agrees with $C_{i'}v$ for a vertex $v \in V_0$ which is contained in two tetrahedra T_i and $T_{i'}$, we have to "stitch" them. What we do is to find a piecewise linear map which is closest to the collection $\{C_i \mid 1 \leq i \leq n\}$ with respect to an *error function*. We discuss different error functions in Sect. 6. The deformed shape $V(w_1, \ldots, w_m)$ is computed as the minimiser of the error function.

In the following sections, we discuss each step in detail.

4 Tetrisation

In computer graphics systems, shapes are usually represented by surface meshes. To apply the ARAP technique described in the previous section, we have to have a tetrahedral structure. Here, we consider a method to build a tetrahedral structure from a given triangular surface mesh.

Definition 2 For a triangular mesh, we denote an element of the vertex set V by a three dimensional vector and an element of the set of (face) triangles $\mathscr{F}$ by an ordered tuple of three vertices (v_1, v_2, v_3). For $(v_1, v_2, v_3) \in \mathscr{F}$, we call the ordered tuples v_1v_2, v_2v_3 and v_3v_1 the *oriented edges*. A triangular mesh is said to be *non-degenerate* when the vertices of each triangle are not co-linear.

Given a triangular mesh, we would like to associate a tetrahedral structure which we can apply the ARAP framework to.

Definition 3 Given a non-degenerate triangular mesh $(V, \mathscr{F})$. A *tetrisation* of $(V, \mathscr{F})$ is a tetrahedral structure which consists of the vertex set $\bar{V}$ and the set of tetrahedra $\mathscr{T}$. We require $(\bar{V}, \mathscr{T})$ to satisfy the following conditions:

1. $V \subset \bar{V}$. That is, $\bar{V}$ is obtained by adding *ghost vertices* to V.
2. Each triangle in $\mathscr{F}$ has to be contained in at least one tetrahedron in $\mathscr{T}$.
3. Each tetrahedron is non-degenerate, that is, the four vertices are not co-planar.

These conditions are exactly what are required in the ARAP framework.

We give three methods to produce tetrisation in the following. Recall that the unit normal vector $n(F)$ of a triangle $F = (v_1, v_2, v_3)$ is computed by $\dfrac{(v_2 - v_1) \times (v_3 - v_1)}{|(v_2 - v_1) \times (v_3 - v_1)|}$, where the denominator $|(v_2 - v_1) \times (v_3 - v_1)|$ is twice the area $2\mathrm{Area}(F)$ of F.

4.1 Face-Normal Tetrisation

We begin with a simple method which has been commonly used in various applications. For each triangle $F = (v_1, v_2, v_3)$ in $\mathscr{F}$, add the ghost vertex

$$v_0 = \frac{(v_1 + v_2 + v_3)}{3} + \frac{(v_2 - v_1) \times (v_3 - v_1)}{\sqrt{|(v_2 - v_1) \times (v_3 - v_1)|}}$$

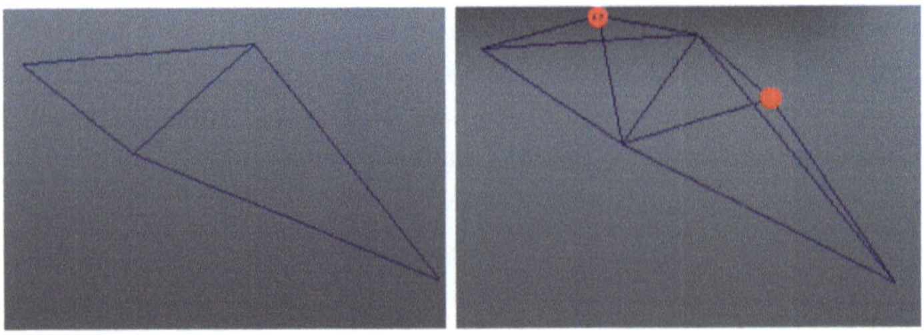

Fig. 3 *Left* the original surface. *Right* its face-normal tetrisation. Ghost vertices are marked with a *red circle*

and form a tetrahedron (v_0, v_1, v_2, v_3). The resulting tetrahedral structure has $\#\mathcal{T} = \#\mathcal{F}$ and $\#\bar{V} = \#V + \#\mathcal{T}$.

A problem with this tetrisation when applied to the ARAP framework is that this does not capture the relation between adjacent triangles. For example, consider two triangles sharing an edge as in Fig. 3. Any rotation invariant error function (see Sect. 6) with $C_1 = C_2 = Id$ will be minimised regardless of the angle between the two triangles. In other words, folds do not cause any penalty in the error function.

4.2 Edge-Normal Tetrisation

We assume each oriented edge appears only once among all the triangles. In other words, an unoriented edge should be contained at most two triangles with opposite orientations. Also, we assume all the triangles have at least one shared edge, that is, there is no "lone" triangle. (We can remove this assumption by adding ghost vertices not only for shared edges but also for all edges. However, this is inefficient and makes no sense.)

For each shared edge v_1v_2, denote by $F_1 = (v_1, v_2, v_3)$ and $F_2 = (v_1, v_4, v_2)$ the two triangles adjacent to it. Add a ghost vertex

$$v_0 = \frac{v_1 + v_2}{2} + |v_1 - v_2| \frac{n(F_1) + n(F_2)}{|n(F_1) + n(F_2)|}$$

and form two tetrahedra (v_0, v_1, v_2, v_3) and (v_0, v_1, v_4, v_2). The resulting tetrahedral structure (Fig. 4) has $\#\mathcal{T} = 2 \cdot \#(\text{shared edges})$ and $\#\bar{V} = \#V + \#(\text{shared edges})$. The idea of this tetrisation is to encode the angle between adjacent triangles, which is neglected by the face-normal tetrisation.

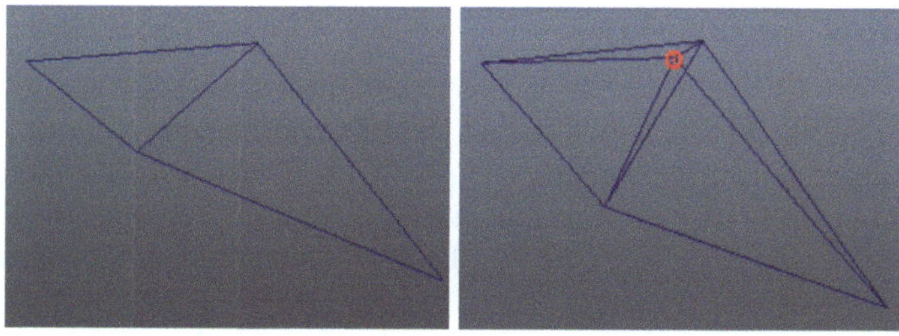

Fig. 4 *Left* the original surface. *Right* its edge-normal tetrisation. Ghost vertex is marked with a *red circle*

4.3 Vertex-Normal Tetrisation

We assume that every vertex has a neighbourhood homeomorphic to the plane or the half plane. In other words, the mesh is a manifold (with boundary). Also, we assume all the triangles have at least one shared vertex. (Again, we can remove this assumption as in the previous subsection.)

For each shared vertex v, denote the adjacent triangles by $F_1, F_2, \ldots,$ and F_k. Add a ghost vertex

$$v_0 = v + \sqrt{\sum_{i=1}^{k} \mathrm{Area}(F_i)}\frac{n(F_1) + \cdots + n(F_k)}{|n(F_1) + \cdots + n(F_k)|}$$

and form k tetrahedra by adding v_0 to the triangles F_i $(1 \leq i \leq k)$. The resulting tetrahedral structure (Fig. 5) has $\#\mathscr{T} = 3\#\mathscr{F} - \#(\text{non-shared vertices})$ and $\#\bar{V} = \#V + \#(\text{shared vertices})$. An advantage of this method is that it extends straightforwardly to general polyhedral meshes. The idea of this tetrisation is to encode the angle around internal vertices, which is neglected by the face-normal tetrisation.

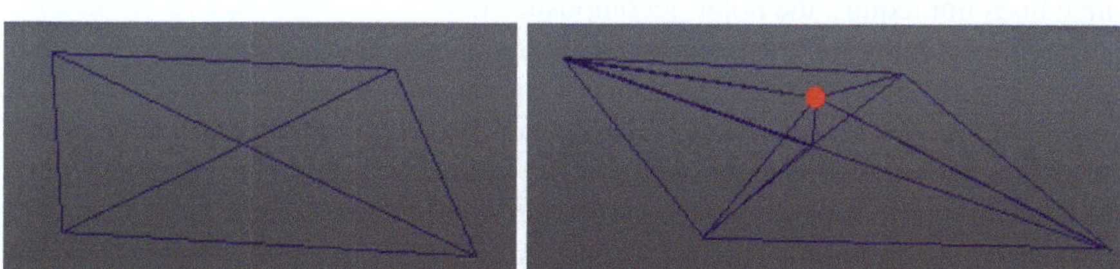

Fig. 5 *Left* the original surface. *Right* its vertex-normal tetrisation. Ghost vertex is marked with a *red circle*. Ghost vertices on the boundary are omitted for simplicity

5 Blending Linear Maps

In this section, we discuss how to blend local transformations $\hat{A}_{1i}, \hat{A}_{2i}, \ldots, \hat{A}_{mi} \in \mathrm{GL}^+(3)$ with regard to the weights $w_1, \ldots, w_m \in \mathbb{R}$ to obtain the blended local transformation $C_i \in \mathrm{GL}^+(3)$. For this purpose, we use a function Blend : $\mathbb{R}^m \times (\mathrm{GL}^+(3))^m \to \mathrm{GL}^+(3)$ which satisfies the obvious requirement for interpolation. Then, we set

$$C_i := \mathrm{Blend}(w_1, \ldots, w_m, \hat{A}_{1i}, \hat{A}_{2i}, \ldots, \hat{A}_{mi}).$$

We investigate two such interpolation functions.

First, decompose each $\hat{A}_{ki}$ by the polar decomposition (see, for example, [5, 12])

$$\hat{A}_{ki} = R_{ki} S_{ki}$$

where $R_{ki} \in \mathrm{SO}(3)$ is the rotation and $S_{ki} \in \mathrm{Sym}^+(3)$ is the shear. In [16], they suggest

$$\mathrm{Blend}_P(w_1, \ldots, w_m, \hat{A}_{1i}, \ldots, \hat{A}_{mi}) = \exp\left(\sum_{k=1}^{m} w_k \log(R_{ki})\right)\left(\sum_{k=1}^{m} w_k S_{ki} + \left(1 - \sum_{k=1}^{m} w_k\right) I\right),$$

where log is the principal matrix logarithm and I is the identity matrix.[1] This coincides with the one used in [1] when $m = 1$. On the other hand, we suggest

$$\mathrm{Blend}_C(w_1, \ldots, w_m, \hat{A}_{1i}, \ldots, \hat{A}_{mi}) = \exp\left(\sum_{k=1}^{m} w_k \log^c(R_{ki})\right)\exp\left(\sum_{k=1}^{m} w_k \log(S_{ki})\right), \tag{3}$$

where $\log^c$ is the "continuous" logarithm such that it chooses the nearest branch of logarithm to the adjacent tetrahedra when i varies (see [8] for details). The indeterminacy of log for SO(3) is in the rotation angle and $\log^c$ chooses the angle continuously for adjacent tetrahedra. Note that [8] provides a direct and fast formula for Blend_C which does not require the polar decomposition.

They look similar but there are two significant differences; blending for the shear part and logarithm for SO(3). The value of Blend_P can fall out of $\mathrm{GL}^+(3)$ due to the linear blending of the shear part, which causes distortion in the output (Fig. 6). The use of the continuous logarithm enables the system to produce a smoother morph among shapes which performs large rotation in between (Fig. 7). Note that in [1] which discusses morphing of two shapes, they suggest to use the quaternions and SLERP ([11]) to interpolate the rotation part and the linear interpolation for the shear part. (With three or more shapes, one can use the linear blending of the quaternions for the rotation part as in [9].) However, this method shows similar deficiency as Blend_P.

[1] The term involving I is for normalisation and it enforces $\mathrm{Blend}_P(0, \ldots, 0, \hat{A}_{1i}, \hat{A}_{2i}, \ldots, \hat{A}_{mi}) = I$.

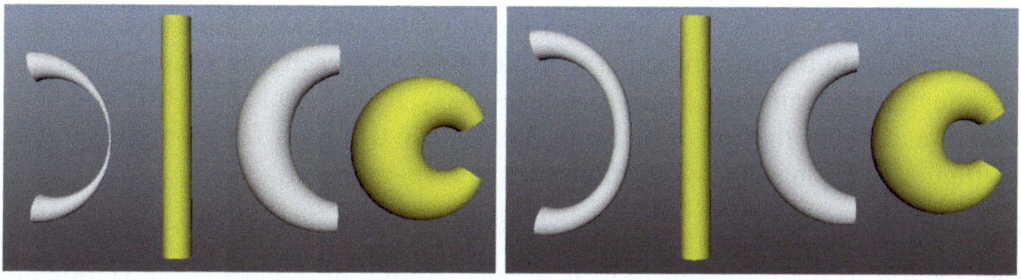

Fig. 6 Interpolation/extrapolation of *yellow shapes*. *Left* with Blend_P function in [16], the extrapolated shape on the *left* is degenerate. *Right* with our Blend_C function, the extrapolated shape is non-degenerate

Fig. 7 Interpolation of *yellow shapes*. *Left* with Blend_P function in [16], some parts try to rotate inconsistently. *Right* with our Blend_C function, local rotations are appropriately handled to produce a smooth interpolation

6 Error Function

In this section, we consider error functions to stitch fragmented tetrahedra. Fix the vertex positions V_0 of the rest shape and the local transformations $\{C_i \mid 1 \leq i \leq n\}$ of the tetrahedra. An error function is a function of the deformed vertex positions V'. By Eq. (2), a piecewise linear map $\{A_i \mid 1 \leq i \leq n\}$ and V' are linearly related and we identify them. In [1], they introduced

$$E_T(V', \{C_i\}) = \sum_{i=1}^{n} |\hat{A}_i - C_i|_F^2 \tag{4}$$

and it has been used in many of the ARAP based shape deformation applications including [2, 15–17]. Note that the function is translation invariant but not rotation invariant. Rotation invariance is sometimes preferable in shape deformation (see, for example, [10, 14] and Fig. 9). We propose an alternative error function which is rotation and translation invariant:

$$E_S(V', \{C_i\}) = \sum_{i=1}^{n} |S(\hat{A}_i) - S(C_i)|_F^2, \tag{5}$$

where $S(X)$ for $X \in \mathrm{GL}(3)$ is the shear factor of the polar decomposition of X (see [12]). Intuitively, this error function measures how much each tetrahedron is distorted. Despite the simplicity and its invariance property, E_S has not been considered in the literature as far as the author is aware. We believe this error function gives a good alternative to E_T in some applications (see Fig. 9).

Remark 2 We can assign a weight $W_i \in \mathbb{R}$ to each tetrahedron T_i to specify its contribution to the error function. It is done simply by replacing the summation $\sum_{i=1}^{n}$ with the weighted one $\sum_{i=1}^{n} W_i$ in the definitions of the error functions. For notational simplicity, we omit them in this paper.

As we described in Sect. 3, we define the output as the minimiser of the error function. In other words, we compute the piecewise linear function $\{A_i \mid 1 \leq i \leq n\}$ which is closest to $\{C_i \mid 1 \leq i \leq n\}$ with respect to the error function. Computing the minimiser for E_T is reduced to solving a sparse linear system (see [1, 16]). For E_S, the computation is not linear. An iterative way similar to [14] is given as follows:

1. Compute the minimiser of $E_T(V', \{C_i\})$ and set $\hat{A}_i$.
2. Compute the polar decomposition $\hat{A}_i = R_i S_i$.
3. Compute the minimiser of $E_T(V', \{R_i S(C_i)\})$ to update $\{\hat{A}_i\}$.
4. Repeat (2) and (3) until $\{\hat{A}_i\}$ converge.

Note that there is some indeterminacy of the minimiser coming from the symmetry of the error function. For example, any translation of a minimiser is also a minimiser. To obtain a unique minimiser, one can impose additional constraints; for E_T fixing the position of the barycentre and for E_S fixing the position of the barycentre and the orientation of some tetrahedra.

7 Implementation

We implemented our algorithm as the Autodesk Maya plugin [6]. In our system, the user can specify the weight for each shape with sliders, or the ball controller which computes the weights by [4] from the configuration of the balls representing the shapes (Fig. 8).

The ARAP framework was also applied to shape blending in [2] in the 2D setting and in [16] in the 3D setting. We demonstrate our improvement discussed in Sects. 4 and 6 by comparing with [16]. First, we note that in [16], (i) the face-normal tetrisation, (ii) the error function E_T, (iii) and the blending function Blend_P are used. We

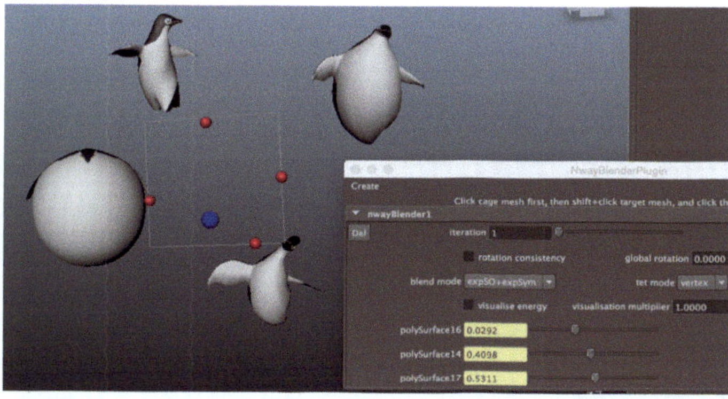

Fig. 8 Our Maya plugin

have already seen the difference between the blending functions Blend_P and Blend_C at the end of Sect. 5. We will turn our attention to (i) and (ii). Figure 9 visually compares different tetrisations in Sect. 4 and the error functions E_T and E_S in Sect. 6. We observe that E_S produces more natural results than E_T but much slower as we see in Table 1. With E_S, the face-normal tetrisation causes extra wrinkles compared to the edge-normal and the vertex-normal tetrisations. As far as we experimented, it depends on the character of shapes to be blended which tetrisation gives the best result. In general, with E_T the output is more or less similar regardless of the choice of tetrisation. With E_S, the vertex-normal tetrisation seems to be a good choice.

Table 1 shows a timing comparison for different tetrisations and error functions. We blended two 3D models each with 26 k triangles on a Macbook Air with 1.7 GHz Intel Core i7 and 8 GB memory. Initialisation part involves the Cholesky decomposition of the space matrix necessary to solve the minimiser of the error functions. This is computed only once in the initialisation process. Note that the matrix is dependent on the tetrahedral structure but independent of the choice of the error function. Runtime part consists of finding the minimiser of the error functions and the computation of Blend functions.

Fig. 9 *Top row* from *left* to *right* rest shape V_0 and its two deformations V_1 and V_2 to be blended with weights $w_1 = 1.0$ and $w_2 = 1.5$. *Second row* from *left* to *right* results obtained by face-normal, edge-normal, and vertex-normal tetrisation with E_T. *Third row* same as the second row but with E_S

Table 1 Timing comparison

	Face E_T	Edge E_T	Vertex E_T	Face E_S	Edge E_S	Vertex E_S
Initialisation (s)	0.1976	0.3080	0.3556	0.1976	0.3080	0.3556
Runtime with Blend_P	45.45 fps	27.43 fps	30.86 fps	11.66 fps	4.132 fps	4.762 fps
Runtime with Blend_C	48.46 fps	29.31 fps	31.46 fps	12.19 fps	4.576 fps	5.037 fps

Acknowledgments This work was partially supported by the Core Research for Evolutional Science and Technology (CREST) Program titled "Mathematics for Computer Graphics" of the Japan Science and Technology Agency (JST), by KAKENHI Grant-in-Aid for Young Scientists (B) 26800043, and by JSPS Postdoctoral Fellowships for Research Abroad.

References

1. M. Alexa, D. Cohen-Or, D. Levin, As-Rigid-As-Possible Shape Interpolation. Proc. ACM SIGGRAPH **2000**, 157–164 (2000)
2. W. Baxter, P. Barla, K. Anjyo, N-way morphing for 2D animation. Comput. Anim. Virtual Worlds **20**(2–3), 79–87 (2009)
3. M. Botsch, O. Sorkine, On linear variational surface deformation methods. IEEE Trans. Vis. Comput. Gr. **14**(1), 213–230 (2008)
4. M.S. Floater, Mean value coordinates. Comput. Aided Geom. Des. **20**(1), 19–27 (2003)
5. N. Higham, Computing the polar decomposition-with applications. SIAM J. Sci. Stat. Comput. **7**(4), 1160–1174 (1986)
6. S. Kaji, An N-way morphing plugin for Autodesk Maya, https://github.com/shizuo-kaji/NWayBlenderMaya
7. S. Kaji, G. Liu, Probe-type deformers, in *Mathematical Progress in Expressive Image Synthesis II* (Springer, Japan, 2015), pp. 63–77
8. S. Kaji, H. Ochiai, A concise parametrisation of affine transformation, preprint arXiv:1507.05290
9. L. Kavan, S. Collins, J. Žára, C. O'Sullivan, Geometric skinning with approximate dual quaternion blending. ACM Trans. Gr. **27**(4), 105:1–105:23 (2008)
10. Y. Lipman, O. Sorkine, D. Levin, D. Cohen-Or, Linear rotation-invariant coordinates for meshes. ACM Trans. Gr. **24**(3), 479–487 (2005)
11. K. Shoemake, Animating rotation with quaternion curves. ACM SIGGRAPH (1985), pp. 245–254
12. K. Shoemake, T. Duff, Matrix animation and polar decomposition, in *Proceedings of Graphics interface '92*, ed. by K.S. Booth, A. Fournier (Morgan Kaufmann Publishers Inc., San Francisco, 1992), pp. 258–264
13. H. Si, TetGen, a Delaunay-based quality tetrahedral mesh generator. ACM Trans. Math. Softw. **41**(2), 11:1–11:36 (2015)
14. O. Sorkine, M. Alexa, As-rigid-as-possible surface modeling, in *Proceedings of Eurographics SGP '07, Eurographics Association, Aire-la-Ville, Switzerland* (2007), pp. 109–116
15. R.W. Sumner, J. Popović, Deformation transfer for triangle meshes. ACM Trans. Gr. **23**(3), 399–405 (2004)
16. R.W. Sumner, M. Zwicker, C. Gotsman, J. Popović, Mesh-based inverse kinematics. ACM Trans. Gr. **24**(3), 488–495 (2005)
17. Y. Yu, K. Zhou, D. Xu, X. Shi, H. Bao, B. Guo, H.-Y. Shum, Mesh editing with poisson-based gradient field manipulation. ACM Trans. Gr. **23**(3), 644–651 (2004)

A Construction Method for Discrete Constant Negative Gaussian Curvature Surfaces

Shimpei Kobayashi

Abstract This article is an application of the author's paper (Kobayashi, Nonlinear d'Alembert formula for discrete pseudospherical surfaces, 2015, [9]) about a construction method for discrete constant negative Gaussian curvature surfaces, the nonlinear d'Alembert formula. The heart of this formula is the Birkhoff decomposition, and we give a simple algorithm for the Birkhoff decomposition in Lemma 3.1. As an application, we draw figures of discrete constant negative Gaussian curvature surfaces given by this method (Figs. 1 and 2).

Keywords Discrete differential geometry · Pseudospherical surface · Loop groups · Integrable systems

1 Introduction

The study of smooth constant negative Gaussian curvature surfaces (PS surfaces[1] in this article) is a classical subject of differential geometry. It is known that the Gauss–Codazzi equations (nonlinear partial differential equations) for a PS surface become a famous integrable system, *sine-Gordon* equation:

$$\partial_y \partial_x u - \sin u = 0.$$

One of the prominent features of integrable systems is that they can be obtained by compatibility conditions for certain linear partial differential equations, the so-called *Lax pair*. Moreover, the Lax pair contains an additional parameter, the *spectral parameter*, and it is a fundamental tool to study integrable systems. On a PS surface, the spectral parameter induces a family of PS surfaces, which will be called the

[1] A constant negative Gaussian curvature surface is sometimes called a pseudospherical surface, thus we use "PS" for the shortened name.

S. Kobayashi (✉)
Department of Mathematics, Hokkaido University, Sapporo 060-0810, Japan
e-mail: shimpei@math.sci.hokudai.ac.jp

Y. Dobashi and H. Ochiai (eds.), *Mathematical Progress in Expressive Image Synthesis III*, Mathematics for Industry 24,
DOI 10.1007/978-981-10-1076-7_3

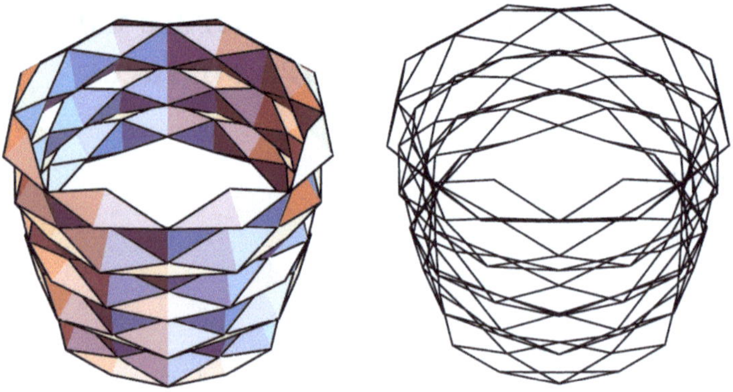

Fig. 1 A discrete PS surface of revolution with the parameters $q = 0.8$ and $c = \pi/8$. The *right figure* is a wired model of the *left* one. The figures are made by using Wolfram Mathematica 10

Fig. 2 (*Left*) a discrete PS surface is given by the same data in Fig. 1 and the spectral parameter has been chosen as $\lambda = 1.5$ in Sym formula (6). The surface has a screw motion symmetry. (*Right*) a discrete PS surface of revolution with the parameters $q = 0.4$ and $c = \pi/8$

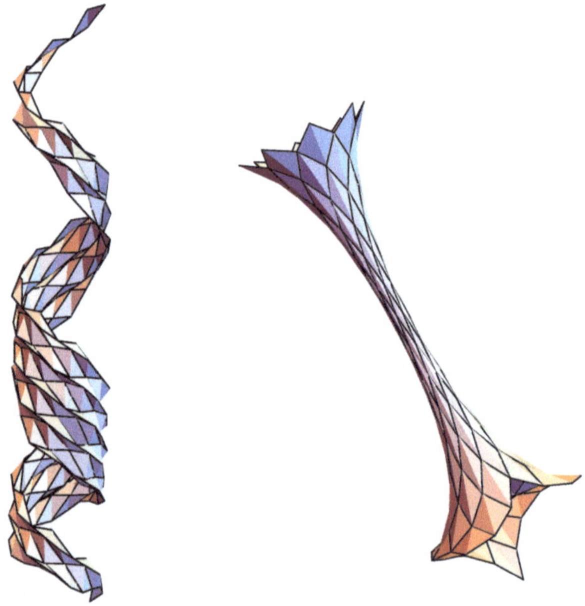

associated family, and the Lax pair is a family of moving frames (Darboux frames) and it will be called the *extended frame* of a PS surface. The extended frame can be thought as an element of the set of maps from the unit circle S^1 in the complex plane into a Lie group, the *loop group*, see Appendix for the definition.

In [10, 13], it was shown that loop group decompositions (Birkhoff decompositions, see Theorem 3.3) of the extended frame F of a PS surface induced a pair of 1-forms (ξ_+, ξ_-), that is, $F = F_+F_- = G_-G_+$ with $\xi_+ = F_+^{-1}dF_+$ and $\xi_- = G_-^{-1}dG_-$. Then it was proved that ξ_+ and ξ_- depended only on x and y, respectively. Conversely it was shown that solving the pair of ordinary differential equations $dF_+ = F_+\xi_+$ and $dG_+ = G_+\xi_-$ and using the loop group decomposition, the extended frame could be recovered. This construction is called the *nonlinear d'Alembert formula* for PS surfaces.

On the one hand a discrete analogue of smooth PS surfaces was defined in [1] and the nonlinear d'Alembert formula for discrete PS surfaces was recently shown in [9]. In this article we first review basic results for smooth/discrete PS surfaces and the nonlinear d'Alembert formula for smooth/discrete PS surfaces according to [1, 5, 9]. The heart of the formula is the Birkhoff decomposition. We next give a simple algorithm (Lemma 3.1) for the Birkhoff decomposition. As an application, we finally draw figures of discrete pseudospherical surfaces given by this method (Figs. 1 and 2).

2 Preliminaries

We briefly recall basic notation and results about smooth and discrete PS surfaces in the Euclidean three space $\mathbb{E}^3$, that is $\mathbb{R}^3$ with the standard inner product $\langle \cdot, \cdot \rangle$, see for examples [1, 5, 10, 11, 13]. Moreover, we recall the nonlinear d'Alembert formula for discrete PS surfaces [9].

2.1 Pseudospherical Surfaces

We first identify $\mathbb{E}^3$ with the Lie algebra of the special unitary group SU_2, which will be denoted by $\mathfrak{su}_2$:

$$ {}^t(x, y, z) \in \mathbb{E}^3 \longleftrightarrow \frac{i}{2} x\sigma_1 - \frac{i}{2} y\sigma_2 + \frac{i}{2} z\sigma_3 \in \mathfrak{su}_2, \tag{1} $$

where σ_j $(j = 1, 2, 3)$ are the Pauli matrices as follows:

$$ \sigma_1 = \begin{pmatrix} 0 & 1 \\ 1 & 0 \end{pmatrix}, \quad \sigma_2 = \begin{pmatrix} 0 & -i \\ i & 0 \end{pmatrix} \quad \text{and} \quad \sigma_3 = \begin{pmatrix} 1 & 0 \\ 0 & -1 \end{pmatrix}. $$

Note that the inner product of $\mathbb{E}^3$ can be computed as $\langle \mathbf{x}, \mathbf{y} \rangle = -2\,\mathrm{trace}(XY)$, where $\mathbf{x}, \mathbf{y} \in \mathbb{E}^3$ and $X, Y \in \mathfrak{su}_2$ are the corresponding matrices in (1). Let f be a PS surface in $\mathbb{E}^3$ with Gaussian curvature $K = -1$. It is known that there exist the *Chebyshev coordinates* $(x, y) \in \mathbb{R}^2$ for f, that is, they are asymptotic coordinates normalized by $|f_x| = |f_y| = 1$. Here the subscripts x and y denote the x- and y-derivatives ∂_x and ∂_y, respectively. Then the first and second fundamental forms for f can be computed as

$$ \mathrm{I} = dx^2 + 2\cos u\, dxdy + dy^2, \quad \mathrm{II} = 2\sin u\, dxdy, $$

where $0 < u < \pi/2$ is the angle between two asymptotic lines. Let

$$e_1 = \frac{1}{2}\sec(u/2)(f_x + f_y), \quad e_2 = \frac{1}{2}\csc(u/2)(f_x - f_y) \text{ and } e_3 = e_1 \times e_3$$

be the Darboux frame rotating on the tangent plane clockwise angle u. Note that it is easy to see that $\{e_1, e_2, e_3\}$ is an orthonormal basis of $\mathbb{E}^3$. Under the identification (1), $\{-\frac{i}{2}\sigma_1, -\frac{i}{2}\sigma_2, -\frac{i}{2}\sigma_3\}$ is an orthonormal basis of $\mathbb{E}^3$, and for a given $F \in \mathrm{SU}_2$ and $x \in \mathfrak{su}_2$, $\mathrm{Ad}(F)(x)(:= FxF^{-1})$ denotes the rotation of x. Thus there exists a $\widetilde{F}$ taking values in SU_2 such that

$$e_1 = -\frac{i}{2}\widetilde{F}\sigma_1\widetilde{F}^{-1}, \quad e_2 = -\frac{i}{2}\widetilde{F}\sigma_2\widetilde{F}^{-1} \text{ and } e_3 = -\frac{i}{2}\widetilde{F}\sigma_3\widetilde{F}^{-1}. \tag{2}$$

Without loss of generality, at some base point $(x_*, y_*) \in \mathbb{R}^2$, we have $\tilde{F}(x_*, y_*) = \mathrm{Id}$. Then there exists a family of frames F parametrized by $\lambda \in \mathbb{R}_+ := \{r \in \mathbb{R} \mid r > 0\}$ satisfying the following system of partial differential equations, see [5] in detail:

$$F_x = FU \text{ and } F_y = FV, \tag{3}$$

where

$$U = \frac{i}{2}\begin{pmatrix} -u_x & \lambda \\ \lambda & u_x \end{pmatrix}, \quad V = -\frac{i}{2}\begin{pmatrix} 0 & \lambda^{-1}e^{iu} \\ \lambda^{-1}e^{-iu} & 0 \end{pmatrix}. \tag{4}$$

The parameter $\lambda \in \mathbb{R}_+$ will be called the *spectral parameter*. We choose F such that

$$F|_{\lambda=1} = \widetilde{F} \text{ and } F|_{(x_*, y_*)} = \mathrm{Id}.$$

The compatibility condition of the system in (3), that is $U_y - V_x + [V, U] = 0$, becomes a version of the sine-Gordon equation:

$$u_{xy} - \sin u = 0. \tag{5}$$

It turns out that the sine-Gordon equation is the Gauss–Codazzi equations for PS surfaces. Thus from the fundamental theorem of surface theory there exists a family of PS surfaces parametrized by the spectral parameter $\lambda \in \mathbb{R}_+$. Then the family of frames F will be called the *extended frame* for f.

From the extended frame F, a family of PS surfaces f^λ, $(\lambda \in \mathbb{R}_+)$ is given by the so-called *Sym formula*, [12]:

$$f^\lambda = \lambda \frac{\partial F}{\partial \lambda} F^{-1}\Big|_{\lambda \in \mathbb{R}_+}. \tag{6}$$

The immersion $f^\lambda|_{\lambda=1}$ is the original PS surface f up to rigid motion. The one-parameter family $\{f^\lambda\}_{\lambda\in\mathbb{R}_+}$ will be called the *associated family* of f.

2.2 Nonlinear d'Alembert Formula

Firstly, we note that the extended frame F of a PS surface f is an element of the *loop group* for SU_2, that is, it is a set of smooth maps from S^1 into SU_2, see Appendix for the definition. In fact by (4) the extended frame is defined on $\mathbb{C}^\times = \mathbb{C}\setminus\{0\}$ and it can be thought as an element in the loop group of SU_2. Then the loop group becomes a Banach Lie group with suitable topology, which is an infinite-dimensional Lie group and thus it will be called the loop group. Then the Birkhoff decomposition of the loop group is fundamental, which will be now explained. The loop group of SU_2 will be denoted by $\Lambda\mathrm{SU}_2$ and we consider two subgroups $\Lambda^+\mathrm{SU}_2$ and $\Lambda^-\mathrm{SU}_2$ of $\Lambda\mathrm{SU}_2$ as sets of maps which can be extended inside the unit disk and outside the unit disk, respectively. In other words, maps $F \in \Lambda\mathrm{SU}_2$, $F_+ \in \Lambda^+\mathrm{SU}_2$ and $F_- \in \Lambda^-\mathrm{SU}_2$ have the following Fourier expansions:

$$F = \sum_{j=-\infty}^{\infty} F_j\lambda^j, \quad F_+ = \sum_{j=0}^{\infty} F_j^+\lambda^j \quad \text{and} \quad F_- = \sum_{j=-\infty}^{0} F_j^-\lambda^j.$$

Then we consider the following problem: for a given map $F \in \Lambda\mathrm{SU}_2$, does there exist $F_\pm$ or $G_\pm$ taking values in $\Lambda^\pm\mathrm{SU}_2$ such that

$$F = F_+F_- \ \text{ or } \ F = G_-G_+$$

holds? The Birkhoff decomposition theorem assures that this decomposition always holds in case of the loop group of SU_2, see Theorem 3.3 in detail. By using the Birkhoff decomposition theorem, we give a construction method for PS surfaces, the so-called *the nonlinear d'Alembert formula.*

From now on, for simplicity, we assume that the base point is $(x_*, y_*) = (0, 0)$ and the extended frame F at the base point is identity:

$$F(0, 0, \lambda) = \mathrm{Id}\,.$$

The nonlinear d'Alembert formula for smooth PS surfaces is summarized as follows, [5, 10, 13].

Theorem 2.1 ([5, 13]) *Let F be the extended frame for a* PS *surface f in $\mathbb{E}^3$. Moreover, let $F = F_+F_- = G_-G_+$ be the Birkhoff decompositions given in Theorem 3.3, respectively. Then F_+ and G_- do not depend on y and x, respectively, and the Maurer–Cartan forms of F_+ and G_- are given as follows*:

$$\begin{cases} \xi_+ = F_+^{-1} dF_+ = \frac{i}{2}\lambda \begin{pmatrix} 0 & e^{-i\alpha(x)} \\ e^{i\alpha(x)} & 0 \end{pmatrix} dx, \\ \xi_- = G_-^{-1} dG_- = -\frac{i}{2}\lambda^{-1} \begin{pmatrix} 0 & e^{i\beta(y)} \\ e^{-i\beta(y)} & 0 \end{pmatrix} dy, \end{cases} \tag{7}$$

where, using the angle function $u(x, y)$, α and β are given by

$$\alpha(x) = u(x, 0) - u(0, 0) \ \ and \ \ \beta(y) = u(0, y).$$

Conversely, let $\xi_\pm$ be a pair of 1-forms defined in (7) with functions $\alpha(x)$ and $\beta(y)$ satisfying $\alpha(0) = 0$. Moreover, let F_+ and G_- be solutions of the pair of the following ordinary differential equations:

$$\begin{cases} dF_+ = F_+ \xi_+, \\ dG_- = G_- \xi_-, \end{cases}$$

with $F_+(x = 0, \lambda) = G_-(y = 0, \lambda) = \mathrm{Id}$. Moreover let $D = \mathrm{diag}(e^{-\frac{i}{2}\alpha}, e^{\frac{i}{2}\alpha})$ and decompose $(F_+ D)^{-1} G_-$ by the Birkhoff decomposition in Theorem 3.3:

$$(F_+ D)^{-1} G_- = V_- V_+^{-1},$$

where $V_- \in \Lambda_*^- \mathrm{SU}_2$ and $V_+ \in \Lambda^+ \mathrm{SU}_2$. Then $F = G_- V_+ = F_+ D V_-$ is the extended frame of some PS *surface in $\mathbb{E}^3$.*

Definition 1 The pair of 1-forms (ξ_+, ξ_-) in (7) will be called the pair of normalized potentials.

Definition 2 In [5], it was shown that the extended frames of PS surfaces can be also constructed from the following pair of 1-forms:

$$\eta^x = \sum_{j=-\infty}^{1} \eta_j^x \lambda^j dx \ \ \text{and} \ \ \eta^y = \sum_{j=-1}^{\infty} \eta_j^y \lambda^j dy, \tag{8}$$

where η_j^x and η_j^y take values in $\mathfrak{su}_2$, and each entry of η_j^x (resp. η_j^y) is smooth on x (resp. y), and $\det \eta_1^x \neq 0$, $\det \eta_{-1}^y \neq 0$. Moreover η_j^x and η_j^y are diagonal (resp. off-diagonal) if j is even (resp. odd). This pair of 1-forms (η^x, η^y) is a generalization of the normalized potentials (ξ_+, ξ_-) in (7) and will be called the *pair of generalized potentials*, see also [4].

2.3 Discrete Pseudospherical Surfaces

Discrete PS surfaces were first defined in [1]. Instead of the smooth coordinates $(x, y) \in \mathbb{R}^2$, we use the quadrilateral lattice $(n, m) \in \mathbb{Z}^2$, that is, all functions depend on the lattice $(n, m) \in \mathbb{Z}^2$. The subscripts 1 and 2 (resp. $\bar{1}$ and $\bar{2}$) denote the forward (resp. backward) lattice points with respect to n and m: For a map $s(n, m)$ of the lattice $(n, m) \in \mathbb{Z}^2$, we define $s_1, s_2, s_{\bar{1}}$ and $s_{\bar{2}}$ by

$$s_1 = s(n+1, m),\ \ s_{\bar{1}} = s(n-1, m),\ \ s_2 = s(n, m+1)\ \text{ and }\ s_{\bar{2}} = s(n, m-1).$$

A discrete PS surface $f : \mathbb{Z}^2 \to \mathbb{E}^3$ was defined by the following two conditions:

1. For each point $f \in \mathbb{E}^3$, there is a plane P such that

$$f,\ f_1,\ f_{\bar{1}},\ f_2,\ f_{\bar{2}} \in P.$$

2. The length of the opposite edge of an elementary quadrilateral are equal:

$$|f_1 - f| = |f_{12} - f_2| = a(n) \neq 0,\ \ |f_2 - f| = |f_{12} - f_1| = b(m) \neq 0.$$

Then the *discrete extended frame* F, which takes values in $\Lambda \mathrm{SU}_2$, of a discrete PS surface f can be defined by the following partial difference system, see [1] and [2, Sect. 3.2]:

$$F_1 = FU \ \text{ and } \ F_2 = FV, \tag{9}$$

where

$$\begin{cases} U = \dfrac{1}{\Delta_+}\begin{pmatrix} e^{-\frac{i}{2}(u_1 - u)} & \frac{i}{2}p\lambda \\ \frac{i}{2}p\lambda & e^{\frac{i}{2}(u_1-u)} \end{pmatrix}, \\[2ex] V = \dfrac{1}{\Delta_-}\begin{pmatrix} 1 & -\frac{i}{2}qe^{\frac{i}{2}(u_2+u)}\lambda^{-1} \\ -\frac{i}{2}qe^{-\frac{i}{2}(u_2+u)}\lambda^{-1} & 1 \end{pmatrix}, \end{cases} \tag{10}$$

with $\Delta_+ = \sqrt{1 + (p/2)^2\lambda^2}$ and $\Delta_- = \sqrt{1 + (q/2)^2\lambda^{-2}}$. Here u is a real function depending on both n and m, and $p \neq 0$ and $q \neq 0$ are real functions depending only on n and m, respectively:

$$u = u(n, m),\ \ p = p(n)\ \text{ and }\ q = q(m).$$

The compatibility condition of the system in (9), that is $VU_2 = UV_1$, gives the so-called *discrete sine-Gordon equation*:

$$\sin\left(\frac{u_{12} - u_1 - u_2 + u}{4}\right) = \frac{pq}{4}\sin\left(\frac{u_{12} + u_1 + u_2 + u}{4}\right). \tag{11}$$

The Eq. (11) was first found by Hirota in [7] and also called the *Hirota equation.*

Remark 2.2 Strictly speaking, the lengths of the edges for a discrete PS surface should be small (less than 1). If the length is big (greater than or equal to 1), then the compatibility condition $VU_2 = UV_1$ gives a discrete analogue of mKdV equation, see [8] in detail. Since this restriction is fundamental for the discrete nonlinear d'Alembert formula, we assume the conditions in (13).

Then the discrete PS surface f can be given by the so-called Sym formula, [1]:

$$f^\lambda = \lambda \frac{\partial F}{\partial \lambda} F^{-1}\bigg|_{\lambda \in \mathbb{R}_+}. \tag{12}$$

The original discrete PS surface f and $f^\lambda|_{\lambda=1}$ are the same surface up to rigid motion. It is easy to see that the map f^λ defined in (12) satisfies two properties of a discrete PS surface and $\{f^\lambda\}_{\lambda\in\mathbb{R}_+}$ gives a family of discrete PS surfaces, see [2, Theorem 3].

2.4 Nonlinear d'Alembert Formula for Discrete PS Surfaces

In this subsection we assume that the base point is $(n_*, m_*) = (0, 0)$ and the discrete extended frame F at the base point is identity:

$$F(0, 0, \lambda) = \mathrm{Id}\,.$$

Moreover, we also assume that the functions p and q in (10) satisfy the inequalities,

$$0 < \left|\frac{p}{2}\right| < 1 \quad \text{and} \quad 0 < \left|\frac{q}{2}\right| < 1, \tag{13}$$

see Remark 2.2. Then the discrete nonlinear d'Alembert formula can be summarized as follows.

Theorem 2.3 ([9]) *Let f be a discrete* PS *surface and F the corresponding discrete extended frame. Decompose F according to the Birkhoff decomposition in Theorem 3.3*:

$$F = F_+ F_- = G_- G_+,$$

where $F_+ \in \Lambda_*^+ \mathrm{SU}_2$, $F_- \in \Lambda^- \mathrm{SU}_2$, $G_- \in \Lambda_*^- \mathrm{SU}_2$ and $G_+ \in \Lambda^+ \mathrm{SU}_2$. Then F_+ and G_- do not depend on $m \in \mathbb{Z}$ and $n \in \mathbb{Z}$, respectively, and the discrete Maurer–Cartan forms of F_+ and G_- are given as follows:

$$\begin{cases} \xi_+ = F_+^{-1}(F_+)_1 = \dfrac{1}{\Delta_+}\begin{pmatrix} 1 & \frac{i}{2} p e^{-i\alpha}\lambda \\ \frac{i}{2} p e^{i\alpha}\lambda & 1 \end{pmatrix}, \\ \xi_- = G_-^{-1}(G_-)_2 = \dfrac{1}{\Delta_-}\begin{pmatrix} 1 & -\frac{i}{2} q e^{i\beta}\lambda^{-1} \\ -\frac{i}{2} q e^{-i\beta}\lambda^{-1} & 1 \end{pmatrix}, \end{cases} \tag{14}$$

where $\Delta_+ = \sqrt{1 + (p/2)^2\lambda^2}$ *and* $\Delta_- = \sqrt{1 + (q/2)^2\lambda^{-2}}$, *the functions* p *and* q *are given in (10), and* α *and* β *are functions of* $n \in \mathbb{Z}$ *and* $m \in \mathbb{Z}$, *respectively. Moreover using the function* $u(n, m)$ *in (10),* $\alpha(n)$ *and* $\beta(m)$ *are given by*

$$\begin{cases} \alpha(n) = \frac{1}{2}u(n+1, 0) + \frac{1}{2}u(n, 0) - u(0, 0), \\ \beta(m) = \frac{1}{2}u(0, m+1) + \frac{1}{2}u(0, m). \end{cases} \tag{15}$$

Conversely, Let $\xi_\pm$ *be a pair of matrices defined in (14) with arbitrary functions* $\alpha = \alpha(n)$, $\beta = \beta(m)$ *with* $\alpha(0) = 0$ *and* $p = p(n)$, $q = q(m)$ *satisfying the conditions (13). Moreover, let* $F_+ = F_+(n, \lambda)$ *and* $G_- = G_-(m, \lambda)$ *be the solutions of the ordinary difference equations*

$$(F_+)_1 = F_+\xi_+ \text{ and } (G_-)_2 = G_-\xi_-, \tag{16}$$

with $F_+(n = 0, \lambda) = G_-(m = 0, \lambda) = \mathrm{Id}$ *and set a matrix* $D = \mathrm{diag}(e^{\frac{i}{2}k}, e^{-\frac{i}{2}k}) \in \mathrm{U}_1$, *where* $k(0) = 0$ *and* $k(n) = 2\sum_{j=0}^{n-1}(-1)^{j+n}\alpha(j)$ *for* $n \geqq 1$. *Decompose* $(F_+D)^{-1}G_-$ *by the Birkhoff decomposition in Theorem 3.3*:

$$(F_+D)^{-1}G_- = V_-V_+^{-1}, \tag{17}$$

where $V_- \in \Lambda_*^-\mathrm{SU}_2$, $V_+ \in \Lambda^+\mathrm{SU}_2$. *Then* $F = G_-V_+ = F_+DV_-$ *is the discrete extended frame of some discrete* PS *surface in* $\mathbb{E}^3$. *Moreover the solution* $u = u(n, m)$ *of the discrete sine-Gordon for the discrete* PS *surface satisfies the relations in (15).*

Definition 3 The pair of matrices (ξ_-, ξ_+) given in (14) will be called the pair of discrete normalized potentials.

Similar to the smooth case, we generalize the pair of discrete normalized potentials:

Definition 4 Let (ξ_-, ξ_+) be a pair of discrete normalized potentials and let η_m and η_n be

$$\eta_n = P_-^l\xi_+P_-^r, \quad \eta_m = P_+^l\xi_-P_+^r. \tag{18}$$

Here we assume that $P_\pm^\star$ ($\star = l$ or r) take values in $\Lambda^\pm\mathrm{SU}_2$ and do not depend on m and n, respectively, that is, $P_-^\star = P_-^\star(n, \lambda)$ and $P_+^\star = P_+^\star(m, \lambda)$. Thus the η_n and η_m do not depend on m and n, respectively:

$$\eta_n = \eta_n(n, \lambda), \quad \eta_m = \eta_m(m, \lambda).$$

The pair (η_n, η_m) given in (18) will be called the pair of discrete generalized potentials.

Remark 2.4 The pair of normalized potentials (ξ_+, ξ_-) and the corresponding pair of discrete generalized potentials (η_n, η_m) in (18) give in general different discrete PS surfaces.

3 Algorithm for Birkhoff Decomposition

In this section, we give a simple algorithm performing the Birkhoff decomposition used in Theorem 2.3.

When one looks at the discrete extended frame F defined in (9), F_+ and G_- defined in (16), one notices that they are given by products of two types of matrices:

$$e_+ = \frac{1}{\sqrt{1+|a|^2\lambda^2}}\begin{pmatrix} e^{i\theta} & a\lambda \\ -\bar{a}\lambda & e^{-i\theta}\end{pmatrix}, \quad e_- = \frac{1}{\sqrt{1+|b|^2\lambda^{-2}}}\begin{pmatrix} e^{i\kappa} & b\lambda^{-1} \\ -\bar{b}\lambda^{-1} & e^{-i\kappa}\end{pmatrix}, \tag{19}$$

where $\theta, \kappa \in \mathbb{R}, a, b \in \mathbb{C}$ and $|a|, |b| < 1$. It is easy to see that $e_\pm$ take values in $\Lambda^\pm \mathrm{SU}_2$, respectively. Two matrices e_+ and e_- do not commute in general, however, the following lemma holds.

Lemma 3.1 *Let $e_\pm$ be matrices in (19). Then there exist matrices $\tilde{e}_\pm$ which take values in $\Lambda^\pm \mathrm{SU}_2$ such that*

$$e_+e_- = \tilde{e}_-\tilde{e}_+$$

holds. In particular $\tilde{e}_\pm$ can be explicitly computed as follows:

$$\tilde{e}_+ = \frac{1}{\sqrt{1+|\tilde{a}|^2\lambda^2}}\begin{pmatrix} e^{i\tilde{\theta}} & \tilde{a}\lambda \\ -\bar{\tilde{a}}\lambda & e^{-i\tilde{\theta}}\end{pmatrix} \quad \text{and} \quad \tilde{e}_- = \frac{1}{\sqrt{1+|\tilde{b}|^2\lambda^{-2}}}\begin{pmatrix} e^{i\tilde{\kappa}} & \tilde{b}\lambda^{-1} \\ -\bar{\tilde{b}}\lambda^{-1} & e^{-i\tilde{\kappa}}\end{pmatrix},$$

where $\tilde{a}, \tilde{b}, \tilde{\theta}$ and $\tilde{\kappa}$ are explicitly chosen by the following equations:

$$\tilde{a} = ae^{-i(\kappa+\tilde{\kappa})}, \ \tilde{b} = be^{i(\theta+\tilde{\theta})} \ \text{ and } \ \tilde{\theta}+\tilde{\kappa} = \theta+\kappa+2\arg(1-a\bar{b}e^{-i(\theta+\kappa)}).$$

Note that $\tilde{e}_\pm$ are not unique and one can always choose $\tilde{\theta} = 0$ or $\tilde{\kappa} = 0$.

Proof It is just a consequence of a direct computation of e_+e_- and $\tilde{e}_-\tilde{e}_+$, respectively.

Using Lemma 3.1 iteratively, we obtain the following algorithm for the Birkhoff decomposition.

Theorem 3.2 *Let F be the discrete extended frame of a* PS *surface. Moreover let F_+, G_- and D be the matrices defined in (16). Then the Birkhoff decompositions for F and $(F_+D)^{-1}G_-$ can be explicitly computed.*

As an example of the above theorem, we draw figures (Figs. 1 and 2) of discrete PS surfaces of revolution according to the following potential, see also [9, Sect. 3]. Let η_n and η_m be $\eta_n = \eta_m^{-1} = A_+LA_-$ with

$$A_\pm = \frac{1}{\Delta_\pm}\begin{pmatrix} 1 & \pm\frac{i}{2}q\lambda^{\pm1} \\ \pm\frac{i}{2}q\lambda^{\pm1} & 1\end{pmatrix} \quad \text{and} \quad L = \mathrm{diag}(e^{ic}, e^{-ic}),$$

where $\Delta_\pm = \sqrt{1+(q/2)^2\lambda^{\pm 2}}$, q $(0 < |q/2| < 1)$ and $c = \pi\ell^{-1}$ $(\ell \in \mathbb{Z}_+)$ are some constants. We note that (η_n, η_m) is a pair of discrete generalized potentials in Definition 4. The pair of solutions for $((F_n)_1, (G_m)_2) = (F_n, G_m)(\eta_n, \eta_m)$ with $F_n(0) = G_m(0) = \mathrm{Id}$ is explicitly given by

$$F_n = (A_+LA_-)^n, \quad \text{and} \quad G_m = (A_+LA_-)^{-m},$$

respectively. Then the Birkhoff decomposition

$$F_n^{-1}G_m = V_-V_+^{-1}$$

is given by using Lemma 3.1. In fact, $F_n^{-1}G_m$ can be rephrased as

$$F_n^{-1}G_m = (A_+LA_-)^{-(n+m)} = \overbrace{(B_-B_+)(B_-B_+)\dots(B_-B_+)}^{n+m},$$

where we set $B_+ = (A_+L)^{-1}$ and $B_- = A_-^{-1}$. Then by using Lemma 3.1, there exist $B_{+,1}$ and $B_{-,1}$ such that $B_+B_- = B_{-,1}B_{+,1}$ holds. Thus we can compute $F_n^{-1}G_m$ as follows:

$$F_n^{-1}G_m = B_-(B_+B_-)\dots(B_+B_-)B_+ = B_-(B_{-,1}B_{+,1})\dots(B_{-,1}B_{+,1})B_+.$$

Next, we use recursively Lemma 3.1 for $B_{-,i}B_{+,i}$ $(i = 1, 2, \dots, n+m-1)$, that is, there exist $B_{+,i+1}$ and $B_{-,i+1}$ such that $B_{+,i}B_{-,i} = B_{-,i+1}B_{+,i+1}$ holds. Finally, $F_n^{-1}G_m$ can be computed as

$$F_n^{-1}G_m = (B_-B_{-,1}\dots B_{-,n+m-1}) \cdot (B_{+,n+m-1}\dots B_{+,1}B_+).$$

Note that since B_- takes values in $\Lambda_*^-\mathrm{SU}_2$, all $B_{-,i}$ also take values in $\Lambda_*^-\mathrm{SU}_2$. Thus $V_- = B_-B_{-,1}\dots B_{-,n+m-1}$ and $V_+^{-1} = B_{+,n+m-1}\dots B_{+,1}B_+$. Then we do not need to compute the diagonal matrix D as in Theorem 2.3 for drawing figures, since any λ-independent diagonal term goes away in Sym formula (12), that is, we can use $F_nV_- = G_mV_+$ in stead of the discrete extended frame F which is given by $F = F_nV_-D = F_nV_-D$.

Acknowledgments The author would like to thank an anonymous referee for helpful comments. The author is partially supported by Kakenhi 26400059.

Appendix

In this appendix we give a definition of the loop group of SU_2 and its subgroups $\Lambda^\pm\mathrm{SU}_2$. Moreover theorem of the Birkhoff decomposition will be stated.

It is easy to see that F defined in (3) together with the condition $F|_{(x_*, y_*)} = \mathrm{Id}$ is an element in the *twisted* SU_2*-loop group*:

$$\Lambda\mathrm{SU}_2 := \left\{ g : \mathbb{R}^\times \cup S^1 \to \mathrm{SL}_2\mathbb{C} \;\middle|\; \begin{array}{l} g \text{ is smooth, } g(\lambda) = \left(\overline{g(\bar{\lambda})}^{-1}\right)^T \\ \text{and } \sigma g(\lambda) = g(-\lambda) \end{array} \right\}, \tag{20}$$

where $\mathbb{R}^\times = \mathbb{R}\backslash\{0\}$, $\sigma X = \mathrm{Ad}(\sigma_3)X = \sigma_3 X \sigma_3^{-1}$, $(X \in \mathrm{SL}_2\mathbb{C})$ is an involution on $\mathrm{SL}_2\mathbb{C}$. In order to make the above group a Banach Lie group, we restrict the occurring matrix coefficients to the Wiener algebra $\mathscr{A} = \{f(\lambda) = \sum_{n\in\mathbb{Z}} f_n\lambda^n : S^1 \to \mathbb{C} \mid \sum_{n\in\mathbb{Z}} |f_n| < \infty\}$, where we denote the Fourier expansion of f on S^1 by $f(\lambda) = \sum_{n\in\mathbb{Z}} f_n\lambda^n$. Then the Wiener algebra is a Banach algebra relative to the norm $\|f\| = \sum |f_n|$ and the loop group $\Lambda\mathrm{SU}_2$ is a Banach Lie group, [6].

Let $\mathbb{D}^+$ and $\mathbb{D}^-$ be the interior of the unit disk in the complex plane and the union of the exterior of the unit disk in the complex plane and infinity, respectively. We first define two subgroups of $\Lambda\mathrm{SU}_2$:

$$\Lambda^+\mathrm{SU}_2 = \left\{g \in \Lambda\mathrm{SU}_2 \mid g \text{ can be analytically extended to } \mathbb{D}^+\right\}, \tag{21}$$

$$\Lambda^-\mathrm{SU}_2 = \left\{g \in \Lambda\mathrm{SU}_2 \mid g \text{ can analytically be extended to } \mathbb{D}^-\right\}. \tag{22}$$

Then $\Lambda_*^+\mathrm{SU}_2$ and $\Lambda_*^-\mathrm{SU}_2$ denote subgroups of $\Lambda^+\mathrm{SU}_2$ and $\Lambda^-\mathrm{SU}_2$ normalized at $\lambda = 0$ and $\lambda = \infty$, respectively:

$$\Lambda_*^+\mathrm{SU}_2 = \left\{g \in \Lambda^+\mathrm{SU}_2 \mid g(\lambda = 0) = \mathrm{Id}\right\},$$

$$\Lambda_*^-\mathrm{SU}_2 = \left\{g \in \Lambda^-\mathrm{SU}_2 \mid g(\lambda = \infty) = \mathrm{Id}\right\}.$$

The following decomposition theorem is fundamental.

Theorem 3.3 (Birkhoff decomposition, [3, 6]) *The multiplication maps*

$$\Lambda_*^+\mathrm{SU}_2 \times \Lambda^-\mathrm{SU}_2 \to \Lambda\mathrm{SU}_2 \;\; \textit{and} \;\; \Lambda_*^-\mathrm{SU}_2 \times \Lambda^+\mathrm{SU}_2 \to \Lambda\mathrm{SU}_2$$

are diffeomorphisms onto $\Lambda\mathrm{SU}_2$*, respectively.*

References

1. A. Bobenko, U. Pinkall, Discrete surfaces with constant negative Gaussian curvature and the Hirota equation. J. Differ. Geom. **43**(3), 527–611 (1996)
2. A. Bobenko, U. Pinkall, Discretization of surfaces and integrable systems. Discrete integrable geometry and physics (Vienna, 1996). Oxf. Lect. Ser. Math. Appl. **16**, 3–58 (1999)
3. D. Brander, Loop group decompositions in almost split real forms and applications to soliton theory and geometry. J. Geom. Phys. **58**(12), 1792–1800 (2008)
4. D. Brander, J. Inoguchi, S.-P. Kobayashi, Constant Gaussian curvature surfaces in the 3-sphere via loop groups. Pac. J. Math. **269**(2), 281–303 (2014)

5. J.F. Dorfmeister, T. Ivey, I. Sterling, Symmetric pseudospherical surfaces. I. General theory. Results Math. **56**(1–4), 3–21 (2009)
6. I.Ts. Gohberg, The factorization problem in normed rings, functions of isometric and symmetric operators, and singular integral equations. Russ. Math. Surv. **19**, 63–114 (1964)
7. R. Hirota, Nonlinear partial difference equations. III. Discrete sine-Gordon equation. J. Phys. Soc. Jpn. **43**(6), 2079–2086 (1977)
8. J. Inoguchi, K. Kajiwara, N. Matsuura, Y. Ohta, Discrete mKdV and discrete sine-Gordon flows on discrete space curves. J. Phys. A **47**(23), 26 (2014) (235022)
9. S.-P. Kobayashi, Nonlinear d'Alembert formula for discrete pseudospherical surfaces. Preprint arXiv:1505.07189 (2015)
10. I.M. Kričever, An analogue of the d'Alembert formula for the equations of a principal chiral field and the sine-Gordon equation. Dokl. Akad. Nauk SSSR **253**(2), 288–292 (1980)
11. M. Melko, I. Sterling, Application of soliton theory to the construction of pseudospherical surfaces in $\mathbf{R}^3$. Ann. Glob. Anal. Geom. **11**(1), 65–107 (1993)
12. A. Sym, Soliton surfaces and their applications (soliton geometry from spectral problems). Geometric aspects of the Einstein equations and integrable systems (Scheveningen, 1984). Lect. Notes Phys. **239**, 154–231 (1985)
13. M. Toda, Weierstrass-type representation of weakly regular pseudospherical surfaces in Euclidean space. Balk. J. Geom. Appl. **7**(2), 87–136 (2002)

Fabrication-Aware Geometry Processing

Daniele Panozzo

Abstract The advent of commodity 3D manufacturing is increasing the demand for advanced design tools that make the designed shapes physically realizable. This is a short overview of fabrication-aware algorithms to solve classical geometry processing problems such as intersection-free mesh deformation, surface parametrization, semi-regular meshing and vector field design.

Keywords Geometry processing · Digital fabrication · Self-supporting surfaces · Hydrographic printing · Quadrilateral meshing · Appearance-mimicking surfaces

The advent of commodity 3D printing is revolutionizing the way people think about designing and prototyping: a designer can now hold in her hands a 3D object hours after its design is complete, drastically reducing costs and enabling quick iterations over many designs. Additive manufacturing enables new applications that were impossible with traditional production processes.

However, the majority of software tools and algorithms currently used to create, manipulate and process digital geometry are not fabrication-aware: they model the shape as an abstract entity that often does not satisfy practical requirements such as stability, robustness or lack of self-intersections. This leads to a large gap between the digital design and the physical fabrication, which is the major obstacle preventing digital fabrication to become mainstream and to deeply change our working habits, in a way similar to the introduction of inkjet and laser printers. This is a short overview of a series of works that strive to fill this gap, providing computational design tools that rely on numerical optimization to create fabrication-ready designs, which can be directly fabricated using digital fabrication technologies.

D. Panozzo (✉)
New York University, 719 Broadway, New York 10003, USA
e-mail: panozzo@cs.nyu.edu

D. Panozzo
ETH Zurich, Universitaetstrasse 6, 8092 Zurich, Switzerland

Y. Dobashi and H. Ochiai (eds.), *Mathematical Progress in Expressive Image Synthesis III*, Mathematics for Industry 24,
DOI 10.1007/978-981-10-1076-7_4

1 Avoiding Self-intersections

A 3D model must be free of self-intersections to be suitable for fabrication. Technically, this means that no elements should intersect (in the case of thin layers such as cloth) and that no volumetric element has negative volume (in case of volumetric representations). This requirement is surprisingly difficult to enforce, especially during design [1], and it is often omitted in favor of simplicity and speed. While this is not a critical problem for models used in movies or games (the overlaps will often not be visible), it becomes mandatory to solve before models can be fabricated. A 3D printer cannot print infinitesimally thin layers, and it thus needs to fill the interior of the shape, which is not defined and impossible to compute in presence of self-intersections (Fig. 1).

Avoiding self-intersections during deformation is particularly simple if each point is restricted to move on a ray pointing the origin [2]. While this reduces considerably the deformation space, this special deformation is ideal to automatically create

Fig. 1 Preventing self-intersections during modeling leads to more intuitive results that are ready to be 3D printed

Fig. 2 A collection of appearance-mimicking surfaces

appearance mimicking surfaces starting from triangle meshes (Fig. 2). The produced surfaces are, by construction, free of self-intersection and ready to be fabricated via 3D printing.

2 Planar Panelization

Quadrilateral meshing algorithms are gaining popularity in the graphics community to convert high resolution triangle meshes into coarse control grids for Catmull–Clark subdivisions. They can also be applied to design planar tessellations, which are ideal for glass and steel construction, due to the low cost of producing flat glass panels. Starting from a planar tessellation, a building can be constructed by replacing each face with a *flat* glass panel, that is much less expensive to manufacture. Mathematically, the edges of a quadrilateral mesh with flat faces define a conjugate field [3], which can be designed with a simple and efficient algorithm [4] that allows architects to interactively experiment with different planar tessellations by simply specifying a set of desired alignment constraints (Fig. 3).

3 Free-Form Masonry Structures

The automatic creation of quadrilateral meshes can be used to design and tessellate of free-form masonry structures [5, 6]. These structures are composed of unsupported stone blocks and they stand thanks to their special geometry where all blocks are in static equilibrium. The block pattern used is a quadrilateral mesh, where an edge every

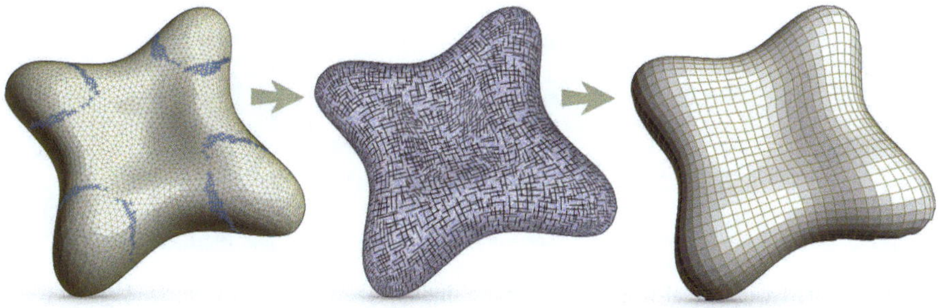

Fig. 3 Planar quadrilateral meshes are used in architectural geometry to design free-form glass and steel structures. The designer specifies a set of alignment constraints (*left*), the constraints are interpolated in a conjugate direction field (*middle*) that is automatically converted into a mesh with planar faces (*right*)

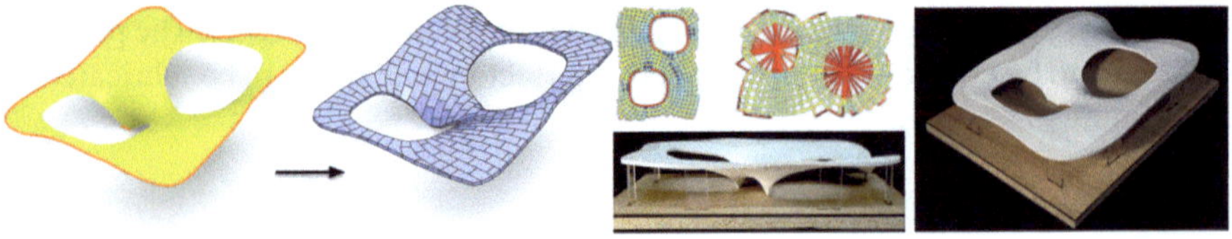

Fig. 4 An input surface is automatically transformed into a masonry 3D model. The equilibrium of the surface is represented by two planar graphs that encode the directions and magnitudes of all forces. The generated blocks are 3D-printed and assembled into a physical model of the surface that stands in compression without using glue or reinforcements

two is removed to create a staggering effect that increases the interlocking between the pieces, simplifying the construction and improving the structural properties of the masonry building (Fig. 4).

4 Hydrographics Printing

In the digital world, assigning arbitrary colors to an object is a simple operation thanks to texture mapping. However, in the real world, the same basic function of applying colors onto an object is far from trivial. One can specify colors during the fabrication process using a color 3D printer, but this does not apply to already existing objects. Paint and decals can be used during post-fabrication, but they are challenging to apply on complex shapes. We proposed a method [7] to enable texture mapping of physical objects, that is, to allow one to map an arbitrary color image onto a three-dimensional object. The approach builds upon hydrographics, a technique to transfer pigments printed on a sheet of polymer onto curved surfaces (Fig. 5).

Fig. 5 We start with a real-world object and a digital 3D model of this object. Using off-the-shelf 3D modeling software, we define a color texture on the digital model. Our algorithm then automatically generates a flat image that we print on a polymer film. We use hydrographics (water transfer printing) to apply this texture onto the real-world object. Our approach compensates for the deformation that happens during the transfer process, so that the final result looks like what we specified on the 3D model

Acknowledgments This work was supported in part by the ERC Starting Grant iModel (StG-2012-306877).

References

1. D. Harmon, D. Panozzo, O. Sorkine, D. Zorin, Interference aware geometric modeling, in *Proceedings of the ACM SIGGRAPH Asia* (2011)
2. C. Schüller, D. Panozzo, O. Sorkine-Hornung, Appearance-mimicking surfaces, in *Proceedings of the ACM SIGGRAPH Asia* (2014)
3. Y. Liu, W. Xu, J. Wang, L. Zhu, B. Guo, F. Chen, G. Wang, General planar quadrilateral mesh design using conjugate direction field, in *Proceedings of the ACM SIGGRAPH Asia* (2011)
4. O. Diamanti, A. Vaxman, D. Panozzo, O. Sorkine-Hornung, Designing N-polyvector fields with complex polynomials, in *Proceedings of the EUROGRAPHICS Symposium on Geometry Processing* (2014)
5. D. Panozzo, P. Block, O. Sorkine-Hornung, Designing unreinforced masonry models, in *Proceedings of the ACM SIGGRAPH* (2013)
6. M. Deuss, D. Panozzo, E. Whiting, Y. Liu, P. Block, O. Sorkine-Hornung, M. Pauly, Assembling self-supporting structures, in *Proceedings of the ACM SIGGRAPH Asia* (2014)
7. D. Panozzo, O. Diamanti, S. Paris, M. Tarini, E. Sorkine, O. Sorkine-Hornung, Texture mapping real-world objects with hydrographics, in *Proceedings of the EUROGRAPHICS Symposium on Geometry Processing* (2015)

Revisiting Vorticity: Pushing Fluid Solvers to the Next Level

Robert Bridson

Abstract Although some of the earliest work in fluid simulation for computer graphics exploited vorticity (e.g. Yaeger et al.'s work on Jupiter for the film 2010 [4]), by and large practical work over the last decade or two has focused on velocity-pressure formulations. This talk looks at why vortex methods are worth coming back to, the troubles that have steered practitioners away from them, and how we might overcome them.

Keywords Fluid simulation · Physics-based animation · Vortex methods

1 Background

From vast oceans to air puffing dust around a footfall, fluids have become a staple of visual effects work, and direct simulation of the underlying physics has emerged as the most attractive approach to generate detailed and natural-looking fluid animation. Current research in graphics on this field can loosely be divided into two parts: work extending the range of phenomena that can be achieved (such as new materials and new means of artist control), and work improving the quality and/or efficiency of standard solves. We focus on the latter in this talk, and mostly on improving smoke simulation in particular.

Measuring the "quality" of a simulation is in of itself a tough, open problem. Accurately solving chaotic, high Reynolds number flow is essentially infeasible, so errors are a given: the question is which sorts of error are acceptable? Backed by experience, I will argue that accurately tracking vorticity and vortex structures is a good goal for graphics.

Vorticity $\boldsymbol{\omega}$ is the curl of velocity $\mathbf{u}$,

$$\boldsymbol{\omega} = \nabla \times \mathbf{u} \tag{1}$$

R. Bridson (✉)
Autodesk Canada, 210 King St. East, Toronto, ON M5A 1J7, Canada
e-mail: robert.bridson@autodesk.com

Y. Dobashi and H. Ochiai (eds.), *Mathematical Progress in Expressive Image Synthesis III*, Mathematics for Industry 24,
DOI 10.1007/978-981-10-1076-7_5

which measures locally how the flow is rotating (as opposed to shearing). In rigidly rotation regions, vorticity is exactly twice the angular velocity of the region. Velocity can be reconstructed from vorticity via the Biot–Savart law, in a sense integrating to undo the differentiation, as long as boundaries are also known: vorticity can serve as the fundamental state of a solver just as well as velocity.

Why vorticity might be so important is best illustrated mathematically with the equations of motion for a constant density fluid in two dimensions with negligible viscosity. Recall the material derivative $D/Dt = \partial/\partial t + \mathbf{u} \cdot \nabla$ tracks how a quantity attached to the fluid changes in time while it flows with the fluid. The usual velocity-pressures equations are

$$\frac{D\mathbf{u}}{Dt} + \frac{1}{\rho}\nabla p = 0, \tag{2}$$

$$\nabla \cdot \mathbf{u} = 0, \tag{3}$$

which show velocity is always being changed by the pressure gradient, while pressure is computed so as to keep the velocity incompressible. On the other hand, the vorticity equation which describes exactly the same motion is this simple:

$$\frac{D\omega}{Dt} = 0. \tag{4}$$

In other words, vorticity just moves with the flow without otherwise changing, which is both striking mathematically and far easier to numerically solve.

Even better, in many flow scenarios vorticity is highly concentrated in small structured regions, and basically is zero throughout most of the domain. This *sparsity* in the representation can also be exploited numerically by tracking vorticity with a small and sparse set of particles or other Lagrangian elements. Figure 1 shows recent smoke results using relatively lightweight vortex triangle meshes and particles.

The fly in the ointment, which I believe has steered people away from vortex methods in the past, is the reconstruction of velocity from vorticity and boundaries. Without boundaries, finding the velocity at a single point in space with the Biot–Savart law requires integrating vorticity with a kernel over the entire fluid domain; with boundaries additional integrals or PDEs are required to calculate their effect. Even just formulating solid and free surface boundary conditions in terms of vorticity can be very tricky indeed. On top of all this, in three dimensions there is an additional term in the vorticity equation, for "vortex stretching," which can be difficult to stably approximate.

Velocity reconstruction needn't be so difficult, however. While a purely Lagrangian approach using the Biot–Savart law may require complex algorithms such as the Fast Multipole Method to scale well, great results can be obtained with the Vortex-in-Cell (VIC) method [2], where vorticity is splatted to a background grid and velocity is reconstructed there via solving the Poisson problem. Even greater detail can be achieved with the Particle-Particle Particle-Mesh (PPPM) approach, without too much more complication for graphics [5].

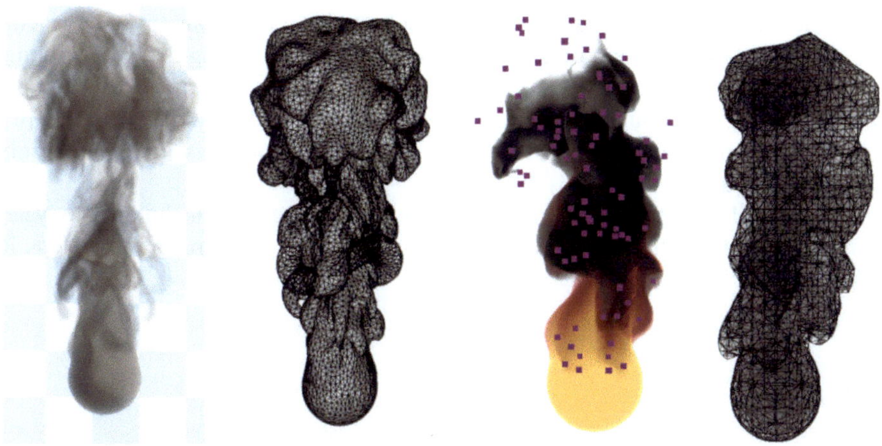

Fig. 1 *Left* Brochu et al. tracked vorticity just on a triangle mesh dividing space between smoky and clear air, producing highly detailed results [1]. *Right* Goldade et al. demonstrated a real-time smoke simulator using a very small number of vortex particles (in *purple*) to nonetheless capture detailed and lively fluid motion [3]

Fig. 2 *Left* Frame from a velocity-pressure smoke simulation using FLIP. *Right* The same with the IVOCK correction to more accurately solve the vorticity equation [6]

While there is still much more to say about boundary conditions for vortex methods, recent work by Zhang et al. shows that it's possible to augment a traditional velocity-pressure solver with a correction to track vorticity like a vortex method, while handling the boundary conditions directly with velocity and pressure [6]. Figure 2 demonstrates the improvement in quality this correction gives with only a small overhead.

There is still much left to do, especially in transferring these techniques to water simulations; I hope this talk will provide a useful view forward on next steps in research.

References

1. T. Brochu, T. Keeler, R. Bridson, Linear-time smoke animation with vortex sheet meshes, in *Proceedings of the ACM SIGGRAPH/Eurographics Symposium on Computer Animation, SCA '12* (2012), pp. 87–95
2. I.P. Christiansen, Numerical simulation of hydrodynamics by the method of point vortices. J. Comp. Phys. **13**(3), 363–379 (1973)
3. R. Goldade, T. Keeler, R. Bridson, Real-time vorticity-based fire and smoke simulation using marching tetrahedra, in *Proceedings of the ACM SIGGRAPH/Eurographics Symposium on Computer Animation* (*Posters*) (2013)
4. L.Yaeger, C. Upson, R. Myers, Combining physical and visual simulation–creation of the planet jupiter for the film, in *Proceedings of the ACM SIGGRAPH* 1986 (2010), pp. 85–93
5. X. Zhang, R. Bridson, A pppm fast summation method for fluids and beyond.ACM Trans. Graph. (Proc. SIGGRAPH Asia) **33**(6), 206:1–206:11 (2014)
6. X. Zhang, R. Bridson, C. Greif, Restoring the missing vorticity in advection-projection fluid solvers. ACM Trans. Graph. (Proc. SIGGRAPH) (2015) (to appear)

Active Comicing for Freehand Drawing Animation

Tsukasa Fukusato and Shigeo Morishima

Abstract This paper presents *Active Comicing*, a prototype sketching system that provides enhanced frame interpolation capability for freehand drawing animation. In this system, the user draws several 2D freeform strokes interactively on multiple frames, and the system automatically constructs stroke-to-stroke interpolation frames. To compose a comprehensive and coherent least-distorting interpolation, we assume input stroke has ghost points, which are additional points defined on stroke edges, and define affine transformations. In addition, the system semi-automatically guides the template motion of each stroke. For example, if the user draws an *arrow*, the system assigns the stroke moves in the direction of the arrow. To assign template motion, we compute the stroke similarity between the user's input and stroke information from a database. With this method, it is possible to generate stroke animation on each frame without stroke interpolation. By combining these techniques, the user can generate freehand animations easily and quickly.

Keywords As-rigid-as-possible stroke interpolation · Stroke matching · Interactive drawing

1 Introduction

2D freehand animation enables viewers to intuitively experience artistry and feeling. Among these techniques, GIF animation (e.g., LINE's stamp and Twitter icon) has attracted worldwide attention in social networks. To present a worldview using 2D freehand animation, anime-like techniques such as flip books and motion comics

T. Fukusato (✉)
Waseda University, 3-4-1 Okubo, Shinjuku-ku, Tokyo 161-8555, Japan
e-mail: tsukasafukusato@gmail.com

S. Morishima
Waseda Research Institute for Science and Engineering, 3-4-1 Okubo, Shinjuku-ku, Tokyo 161-8555, Japan
e-mail: shigeo@waseda.jp

Y. Dobashi and H. Ochiai (eds.), *Mathematical Progress in Expressive Image Synthesis III*, Mathematics for Industry 24,
DOI 10.1007/978-981-10-1076-7_6

are employed. However, the creation of 2D freehand animations has always been a time-consuming and skill-demanding process. Software such as *Adobe Photoshop* provides some assistance by creating animations from a small number of key frames and generating in-between frames automatically. For example, animation software might deform a shape using handles or transform simple geometric primitives. However, creating the numerous key-frames of animation, such as those in a flip book, requires significant skill and time.

Conversely, the cartoon animation industry tends to shift traditional hand-drawn techniques to a pipeline using parameterized 3D models. Although a 3D model technique reduces production costs, this approach comes at the expense of well-established cartoon animation values, such as character and expression. These models may diminish freedom, expressiveness, and the artist's commitment to the characters. It is difficult to parameterize the freedoms of pencil and paper. In short, many amateur animators, including children, find it difficult to create freehand drawing animation.

Our goal is to create new shape from freehand drawing 2D shapes, sketched onto a drawing interface. In this paper, we present Active Comicing, a sketch-based interface that allows users to interpolate freehand drawing strokes without a skeleton (i.e., graffiti animation). Figure 1 shows a freehand stroke animation on a display-integrated tablet. Specifically, we have implemented a freehand stroke interpolation method based on *As-Rigid-As-Possible Stroke Interpolation,* which does not require editing commands or special interaction modes. To address the vertex correspondence problem, in which correspondence between input strokes must be established, our system constructs a simple layer structure. To reduce the distortion of the in-between shapes, we compute the orientation information for stroke vertices. Moreover, this system allows users to edit the animation path of each stroke using a template motion database. As a result, we can easily create a simple 2D freehand drawing animation.

The reminder of this paper is organized as follows. Related works are reviewed in Sect. 2. We discuss the user interface in Sect. 3, and describe the main ideas underlying the proposed method's algorithms in Sect. 4. In Sect. 5, we describe the

Fig. 1 Prototype drawing system with Active Comicing, on a display-integrated tablet

implementation details of our prototype system. We conclude the paper and discuss limitations and future work in Sect. 6.

2 Related Works

Recently, 3D computer graphics researchers have proposed generating 3D models from 2D drawings. In particular, Teddy [1] inflates the stroke region surrounded by a silhouette. River et al. [2] and Yeh et al. [3] propose a method to create 2.5D models, which are hand-drawn illustration style models. Although their method can automatically estimate the depth information of each polygon section, users must configure Z-ordering and create each section from scratch. It is difficult to parameterize freehand drawings created with pencil and paper.

Drawing is a simple tool that reflects the artist's creative sense. Pencil lines or brush coloring can express rich emotions or subtle charms. Live2D [4] can create rich animations based on standard linear interpolation (point-to-point interpolation) while keeping the original charms intact. However, it is necessary to determine the character pose on the frame or redraw some or all of the strokes in the model manually. In addition, significant time is required to create a mesh structure and edit mesh deformations. In image morphing research, Cambell [5] and Baxter [6] propose intuitive approaches, whereby line drawings are interpolated in a pose-space with reduced dimensions. This method enables the subspace of a pose to be browsed. Unfortunately, it is limited to line drawings with the same number of lines, and may give unnatural results because curves are linearly interpolated.

While physically based simulations [7, 8] can also be used for this purpose, it tends to be slow and produces unstable shapes. Wang's approach [9] enables image deformation based on meshless rigid shape matching [10, 11]. Moreover, Sykora et al. [12] apply this approach to elasticity-inspired character registration. With this method, it is possible to register images undergoing large free-form deformations and appearance variations. Unfortunately, they cannot directly obtain pixel or sub-pixel precision, because they embed the image into a coarse lattice. Although this method can apply a multi-scale extension, increasing the number of squares makes the overall iterative process ineffective.

In shape interpolation research, *As-Rigid-As-Possible Shape Interpolation* approaches (ARAP) have been studied [13–19]. These methods enable the volume of the stroke's interior to be maintained and produce more plausible animations by using triangle mesh structures. Furthermore, Baxter [20] has extended this method to examples-way rigid interpolation. However, this system uses polygonal boundary-based triangulation; that is, it focuses only on similarity shape morphing. In addition, it is difficult to edit the mapping of each stroke. Alexa [21] proposes Laplacian coordinate for shape interpolation; however, the morphing results have shrinkage. Sederberg [22] exploits intrinsic blending on a basis of interpolating the respective vertex angles and edge length. Moreover, Whited [23] develops BetweenIT, a technique for stroke interpolation from two key frames. This technique combines stroke motions constructed from logarithmic spiral vertex trajectories with stroke deformations based on curvature averaging and twisting warps. This system provides a context in which the user can guide the system in a natural manner to produce quality results efficiently. However, this system only focuses on tight in-betweens, which are drawn between two key frames that are very similar in shape.

Other approaches have processed more general shapes by considering deformations of a template model. For example, Igarashi's Spatial Keyframing [24] animates 3D objects composed of skeletons. Moreover, applying motion capture data to a single character image based on a skeleton has been studied [25, 26]. However, the range of deformation is limited with these approaches. In addition, these approaches do not specify how handles should be interpolated to achieve plausible interpolations. In contrast, Sumner's [27] Mesh Inverse Kinematics system interpolates between multiple meshes. However, a non-linear inverse kinematics approach is not browsed directly.

To summarize, previous animation techniques and tools have restrictions on the types of input strokes that can be used for similarity stroke morphing, and it is necessary to redraw some or all of the strokes in the model. Therefore, we propose a method to interpolate freehand-drawing strokes. The proposed method is of great value, and can create a simple animation interactively.

3 User Interface

Active Comicing's physical user interface utilizes traditional 2D input devices such as a standard mouse and pen tablet. Figure 2 shows an overview of our system. In drawing mode, a user draws several 2D freeform strokes interactively on some key frames. This system has various user drawing functions such as image (e.g., jpg image file) loading function and key frame copy function. The user can also redraw some or all of the strokes on each frame. Moreover, using onion skinning, the user can make decisions on how to create key frames based on the previous key frames in the sequence.

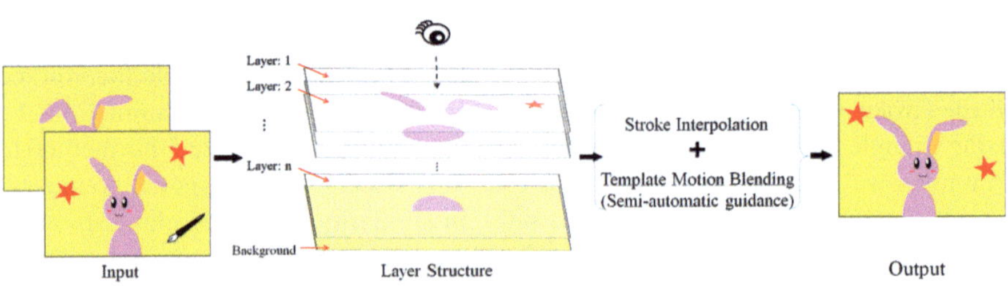

Fig. 2 System overview

A stroke consists of a sequence of points on the plane, which we call stoke vertices. The stroke vertices are interpolated using a centripetal Catmull–Rom spline. In editing mode, the user can transform the strokes to edit the xy-coordinate of each stroke vertex; this translation is performed by dragging the mouse. The system automatically assigns labeling numbers (or layer number) to the strokes on each key frame based on the stroke order. The input strokes on one frame spatially correspond to those on another frame based on the labeling number; stroke-to-stroke correspondences are defined. Moreover, the user can edit the layering order by dragging and dropping layers with the mouse.

The user can easily generate a freehand stroke animation as a GIF image, using the provided animation timeline in animation mode.

4 Algorithm

4.1 Stroke Interpolation Method

To interpolate two frames, the corresponding strokes have to have the same number of stroke vertices. The source strokes are first resampled to the number of target stoke vertices n equidistantly. Let $P = (\mathbf{p}_0, \ldots, \mathbf{p}_n)$ be the source stroke and $Q = (\mathbf{q}_0, \ldots, \mathbf{q}_n)$ be the target stroke.

For 2D interpolation technique, Sederberg [28] proposes a solution to the vertex correspondence problem, and the vertex path problem is dealt with in Sederberg's [22] method, which interpolates the edge lengths and the angles between consecutive edges of polygonal curves. To ensure these blended strokes are closed without local self-intersection, they set to an equality constraint of the end positions by tweaking the edge length only; however, the final morphing results are dependent on the computation order of dihedral angles and edge length. Moreover, they cannot add some constraints, and extend this method to an invariant interpolation under similarity transformation, i.e., rotation and scale. Most shape interpolation and deformation studies have focused on 2D or 3D triangle and lattice [29, 30] because the affine transformation of each triangle polygon can be computed easily. However, this approaches do not determine a vertex path for stroke interpolation. Baxter et al. [20] apply a Delaunay triangulation to 2D stroke vertices, and then deform the Delaunay triangles based on ARAP. Unfortunately, the Delaunay triangulation approach is focused only on a closed stroke, and is less intuitive for interpolations between closed strokes. Specifically, it is difficult to define an affine transformation based on the source and the target stroke vertices only.

Therefore, we assume that each stroke vertex $\mathbf{v}_i$ has a ghost vertex $\mathbf{v}_i^g \in \mathbb{R}^2$, which is placed on a certain distance along normal direction of each adjacent edge (as shown in Fig. 3). Representing the ghost vertex is mainly inspired by Umetani's ghost point approach [31] and Sumner's surface tetrahedra [32]. Using the ghost vertex, we generate the triangles of the source and the target strokes in order to

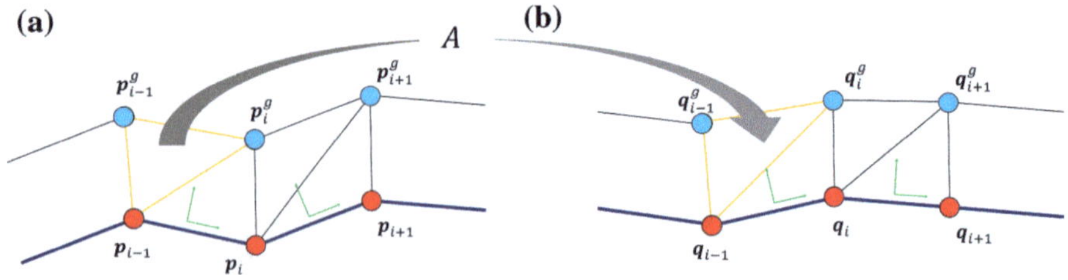

Fig. 3 Configuration of stroke, vertex (*red*) and ghost vertex (*cyan*)

compute a unique affine transformation of each edge. For the corresponding triangle in each stroke shape, the system first computes the ghost vertex $\mathbf{v}^g$ of each stroke vertex (R_{90} denotes rotation matrix by 90°):

$$\mathbf{v}_i^g = \mathbf{v}_i + R_{90}\left(\frac{\mathbf{v}_{i+1} - \mathbf{v}_{i-1}}{2}\right) \tag{1}$$

$$\text{where} \quad R_{90} = \begin{pmatrix} 0 & -1 \\ 1 & 0 \end{pmatrix}$$

As the result, the source and the target stroke can consist of a chain of triangles. Then, we focus on ARAP interpolation of local and global linear transformation. An affine mapping represented by matrix A transforms the source into the target triangle. The matrix A is parameterized by time $t \in \mathbb{R}$ such that $A(0.0) = I$ (identity matrix) and $A(1.0) = A$. We next deal with the 2D interpolation of the entire input strokes (the source and the target triangles). To compute a global transformation $B_i(t)$ based on local translation $A_i(t)$, we use Kaji's local error function using the polar decomposition and the exponential map [16] as follows:

$$A_i(t) = R_\theta^t \cdot \exp(t \log S) \tag{2}$$

$$E_i^R(A_i(t), B_i(t)) = \min_{s,\delta \in \mathbb{R}} \|sR_\delta A_i(t) - B_i(t)\|_F^2 \tag{3}$$

where R_δ is a rotation matrix, and s is scale value. This equation measures how different $A_i(t)$ and $B_i(t)$ are as affine transformations of ith triangle. With the local error functions for each triangle, we combine them into a single global error function. If a local affine transformation can be formed by reflections, we exclude an error value of local triangle distortion from the global error function. The global error function is a positive definite quadratic form. Instead, in-between parameter t requires the solution of a linear system of equations.

In addition, using the ghost vertex, we can unify the stroke's global orientation into a counterclockwise orientation. We compute a sign area S as follows:

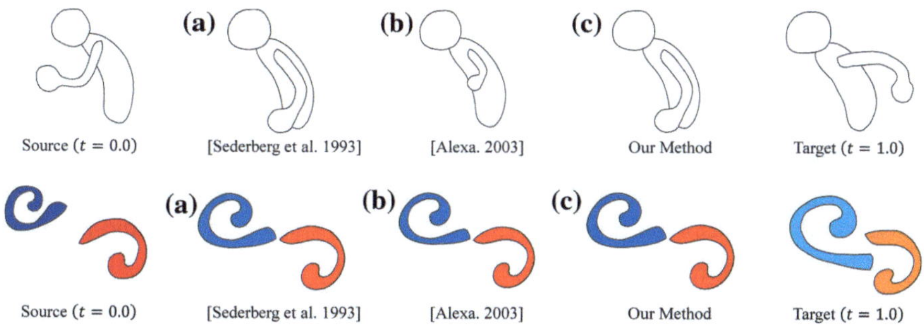

Fig. 4 Comparison. **a** Intrinsic shape interpolation [22] with $t = 0.5$. **b** Laplacian morphing [21]. **c** Our method

$$S = \frac{1}{2}\sum_{i}(v_{ix}v_{(i+1)y} - v_{(i+1)x}v_{iy}) \tag{4}$$

This computation gives a positive signed area S for a simple stroke (non-self-intersecting polygon) ($S > 0$) when the vertices are oriented counterclockwise around the polygon, and negative ($S < 0$) when oriented clockwise. However, this equation cannot be applied to complex strokes (self-intersecting polygon). It is necessary to split the complex strokes into several simple strokes.

For interpolating the line weight and RGBA color information of each stroke, we use standard linear interpolation. The system allows us to set the in-between parameter t for linear and interactive interpolation of each stroke shape. The acceptability of the computation time depends on the shape and the desired application. Figure 4 illustrates the resulting transformations from a source to a target shape. For comparison, Fig. 4a shows Sederberg's method [22] of each vertex coordinate with $t = 0.5$, and Fig. 4b shows Laplacian morphing method [21]. Our transformation (ARAP with ghost vertex) is depicted in Fig. 4c. The results show that we have successfully reduced distortions in stroke shape transformations. In addition, we can incorporate some constraints into the global error function.

4.2 *Template Motion Blending*

In this section, we describe a method to animate strokes based on template effects, such as those in *Microsoft PowerPoint*. The template effects consist of affine transformations (e.g., translation matrix T, scaling matrix S, and rotation matrix R) and alpha blending. This system is formed with the origin at the centroid vertices of each stroke. By setting the stroke motion matrix (e.g., the animation path), we can interactively edit the results of the stroke animation.

In addition, we attempt to synthesize the template effects automatically. For example, when the user draws an *arrow* shape, our system assigns the stroke moves in the direction of the arrow. First, we assume that the user assigns the same effect to similar shape strokes. Therefore, we propose a technique to compute the similarity between strokes, and recommend optimum template effects based on the stroke database, which contains sets of stroke and template effect. We compute the degree of similarity between the input stroke $\mathbf{v}$ and the database stroke $\mathbf{d}$, and present template effects of highly similar strokes from the database. In image processing research, stroke similarity has been proposed [33–36]. Ip et al. [33] exploit affine-invariant stroke features based on stroke area and angle information. Because a single shape signature (in the form of a nineteenth-dimension histogram) records the stroke area and angle information, their similarity between stroke shapes can be computed efficiently using a signature difference. However, this approach, known as histogram intersection, does not consider stroke's global orientation, and has difficulty representing complex stroke shapes. Therefore, to compute stroke similarities, we use rigid shape matching [9, 10] based on the centroid position of the input stroke $\mathbf{v}_{cm}$ and the database stroke $\mathbf{d}_{cm}$. We define the quadratic error function between the stroke vertices $\mathbf{p}_i(= \mathbf{v}_i - \mathbf{v}_{cm})$ and $\mathbf{q}_i(= \mathbf{d}_i - \mathbf{d}_{cm})$ as follows:

$$E = \sum_i |\mathbf{p}_i - sR \cdot \mathbf{q}_i|^2 \tag{5}$$

$$R = A_{pq}S^{-1} = A_{pq}\left(\sqrt{A_{pq}^T A_{pq}}\right)^{-1}$$

$$A_{pq} = \sum_i \mathbf{p}_i \cdot \mathbf{q}_i^T$$

where n is the number of stroke vertices, s is the normalized value ($s = \sum_i |\mathbf{p}_i|/|\mathbf{q}_i|$). The optimal rotation R is the rotation portion of $A_{pq} = RS$; we compute rotation matrix $R = A_{pq}S^{-1}$, where the symmetric portion is $S = \sqrt{A_{pq}^T A_{pq}}$. The output value E provides the dissimilarity value because the value tends to be smaller if two signatures are more similar. However, it is essential to work to have the same number of vertices in a stroke. In this paper, the database strokes are resampled to the number of input stoke vertices during pre-processing.

To evaluate the performance of our similarity for stroke retrieval, we perform an experiment. The experiment is carried out to retrieve relevant strokes based on the users' sketching of the desired stroke with a stroke database of 20 strokes. The evaluation of the approach is based on retrieval accuracy (precision rate). The stroke retrieval results are shown in Table 1. For comparison, we use Ip's affine-invariant histogram approach [33]. These results show that our similarity can determine highly accurate animation template motions. Moreover, we add an editing function for relearning moving guidance. By adding an editing data (a set of stroke and template motion) to the stroke databases, it is possible to obtain a more suitable optimum stroke motion.

Table 1 Stroke classification results

Number of strokes	Our approach (%)	Ip et al. 2002 (%)
10	90.0	50.0
15	75.0	40.0
20	70.0	35.0

5 Implementation

Our prototype system is written using openFrameworks, an open source C++ toolkit. A 64bit Windows PC (Intel®Core^{TM} i7-3770 CPU@3.40 GHz 8 GB RAM; NVIDIA GeForce GT 620M 1 GB) is used. By drawing freehand strokes on some frames ($max\ number$ = 4), users can easily generate animation. Although all results are generated at over 30.0 fps, it is difficult to accurately measure the performance of each computation. Our results are presented in Figs. 5 and 6.

The display-integrated tablet version of Active Comicing has been used to create different 2D freehand animations, mainly by computer graphics researchers and students. Our prototype system was also evaluated by users who provided individual feedback: One user stated that the hand drawing animation capabilities were more impressive and expressive than normal CG animation techniques. Other users commented that they wanted to upload information results to LINE or Twitter, and that the system could benefit from a more elastic function for editing stroke shapes. In the

Fig. 5 Selected examples of deformable stroke animation, 'muscle training', produced by our technique. **a** $t = 0.0$. **b** $t = 0.25$. **c** $t = 0.5$. **d** $t = 0.75$. **e** $t = 1.0$

Fig. 6 Selected examples of deformable stroke animation, 'girl's looking back motion', produced by our technique. **a** $t = 0.0$. **b** $t = 0.25$. **c** $t = 0.5$. **d** $t = 1.0$

future, we plan to include shape deformation functions such as physical simulation in the user interface to create richer animations.

6 Conclusions and Future Works

We have presented a method to interpolate and animate freehand drawing strokes. The prototype system, *Active Comicing,* enables the easy creation of simple 2D GIF animations. Moreover, the results of the stroke animation can be edited according to user preferences by template effects. It is assumed that the stroke similarity technique could also be applied to character recognition to help users find and review required freehand information. We intend to apply the proposed approach to character recognition.

In future work, we plan to increase the number of key-frame, e.g., multi-stroke morphing, and focus on color interpolation, e.g., color model or gradient color. Moreover, we recognize that sampling technique based on stroke shape will help users to create richer animations more efficiently. Therefore, we intend to study these functions. Such functions are applicable to a wide range of situations in anime production.

Acknowledgments This project was funded in part by grants from the Japanese Information-Technology Promotion Agency (IPA) and by Research Fellowship for Young Scientists of Japan Society for the Promotion of Science (JSPS).

References

1. T. Igarashi, S. Matsuoka, H. Tanaka, Teddy: a sketching interface for 3D freeform design, in *Proceedings of the ACM SIGGRAPH 2007 courses*, **21** (ACM, 2007)
2. A. Rivers, T. Igarashi, F. Durand, 2.5D cartoon models. ACM Trans. Graph. (TOG) **29**(4), 59 (2010)
3. C.-K. Yeh, P. Song, P.-Y. Lin, C.-W. Fu, C.-H. Lin, T.-Y. Lee, Double-sided 2.5D graphics. IEEE Trans. Vis. Comput. Graph. (TVCG) **19**(2), 225–245 (2013)
4. Live2D, http://www.live2d.com/en/
5. N.D. Campbell, J. Kautz, Learning a manifold of fonts. ACM Trans. Graph. (TOG) **33**(4), 91 (2014)
6. W. Baxter, K. Anjyo, Latent doodle space. Comput. Graph. Forum (Proc. Eurographics) **25**(3), 477–485 (2006)
7. W.-C. Ma, Y.-H. Wang, G. Fyffe, B.-Y. Chen, P. Debevec, A blendshape model that incorporates physical interaction. Comput. Animat. Virtual Worlds **23**(3–4), 235–243 (2012)
8. S. Setaluri, Y. Wang, N. Mitchell, L. Kavan, Fast grid-based nonlinear elasticity for 2D deformations, in *Proceedings of the 13th ACM SIGGRAPH/Eurographics Symposium on Computer animation* (2014), pp. 67–76
9. Y. Wang, K. Xu, Y. Xiong, Z.-Q. Cheng, 2D shape deformation based on rigid square matching. Comput. Animat. Virtual Worlds **19**(3–4), 411–420 (2008)
10. M. Müller, B. Heidelberger, M. Teschner, M. Gross, Meshless deformations based on shape matching. ACM Trans. Graph. (TOG) **24**(3), 471–478 (2005)

11. M. Müller, N. Chentanez, Solid simulation with oriented particles. ACM Trans. Graph. (TOG) **30**(4), 92 (2011)
12. D. Sỳkora, J. Dingliana, S. Collins, As-rigid-as-possible image registration for hand-drawn cartoon animations, in *Proceedings of the 7th International Symposium on Non-Photorealistic Animation and Rendering* (ACM, 2009), pp. 25–33
13. M. Alexa, D. Cohen-Or, D. Levin, As-rigid-as-possible shape interpolation, in *Proceedings of the 27th Annual Conference on Computer Graphics and Interactive Techniques* (ACM Press/Addison-Wesley Publishing Co., 2001), pp. 157–164
14. T. Igarashi, T. Moscovich, J.F. Hughes, As-rigid-as-possible shape manipulation. ACM Trans. Graph. (TOG) **24**(3), 1134–1141 (2005)
15. W. Baxter, K. Anjyo, Rigid shape interpolation using normal equations, in *Proceedings of the 6th International Symposium on Non-photorealistic Animation and Rendering*(ACM, 2008), pp. 59–64
16. S. Kaji, S. Hirose, S. Sakata, Y. Mizoguchi, K. Anjyo, Mathematical analysis on affine maps for 2D shape interpolation, in *Proceedings of the 11th ACM SIGGRAPH/Eurographics Conference on Computer Animation* (ACM, 2012), pp. 71–76
17. R. Chen, O. Keren, M. Ben-Chen, Planar shape interpolation with bounded distortion. ACM Trans. Graph. (TOG) **32**(4), 108 (2013)
18. H. Ochiai, K. Anjyo, Mathematical basics of motion and deformation in computer graphics, in *Proceedings of the ACM SIGGRAPH 2014 Courses* (ACM, 2014), pp. 19:1–19:47
19. Z. Levi, C. Gotsman, Smooth rotation enhanced as-rigid-as-possible mesh animation. IEEE Trans. Vis. Comput. Graph. (TVCG) **21**(2), 264–277 (2015)
20. W. Baxter, P. Barla, K. Anjyo, N-way morphing for 2D animation. Comput. Animat. Virtual Worlds **20**(2–3), 79–87 (2009)
21. M. Alexa, Differential coordinates for local mesh morphing and deformation. Vis. Comput. **19**(2), 105–114 (2003)
22. T.W. Sederberg, P. Gao, G. Guijin, H. Mu, 2-D shape blending: an intrinsic solution to the vertex path problem, in *Proceedings of the 20th Annual Conference on Computer Graphics and Interactive Techniques* (1993), pp. 15–18
23. B. Whited, G. Noris, M. Simmons, R.W. Robert, M. Gross, J. Rossignac, BetweenIT: an interactive tool for tight inbetweening. Comput. Graph. Forum (Proc. Eurographics) **29**(2), 605–614 (2010)
24. T. Igarashi, T. Moscovich, J.F. Hughes, Spatial keyframing for performance-driven animation, in *Proceedings of the 2005 ACM SIGGRAPH/Eurographics Symposium on Computer Animation* (ACM, 2005), pp. 107–115
25. A. Hornung, E. Dekkers, L. Kobbelt, Character animation from 2D pictures and 3D motion data. ACM Trans. Graph. (TOG) **26**(1), 1 (2007)
26. J. Pan, J.J. Zhang, Sketch-Based Skeleton-Driven 2d Animation and Motion Capture, *Transactions on Edutainment VI* (Springer, Berlin, 2011)
27. R.W. Sumner, M. Zwicker, S.C. Gotsman, J. Popovic, Mesh-based inverse kinematics. ACM Trans. Graph. (TOG) **24**(3), 488–495 (2005)
28. T.W. Sederberg, E. Greenwood, A physically based approach to 2-D shape blending, in *Proceedings of the 19th Annual Conference on Computer Graphics and Interactive Techniques* (1992), 25–34
29. S. Schaefer, T. McPhail, J. Warren, Image deformation using moving least squares. ACM Trans. Graph. (TOG) **25**(3), 533–540 (2006)
30. A. Jacobson, I. Baran, J. Popovic, O. Sorkine, Bounded biharmonic weights for real-time deformation. ACM Trans. Graph. (TOG) **30**(4), 78 (2011)
31. N. Umetani, R. Schmidt, J. Stam, Position-based elastic rod, in *Proceedings of the 13rd ACM SIGGRAPH/Eurographics Symposium on Computer Animation* (2014), pp. 21–30
32. R.W. Sumner, J. Popović, Deformation transfer for triangle meshes. ACM Trans. Graph. (TOG) **23**(3), 399–405 (2004)
33. H.H.-S. Ip, A.K. Cheng, W.Y. Wong, Affine invariant retrieval of shapes based on hand-drawn sketches, in *Proceedings of the 16th International Conference on Pattern Recognition* (IEEE, 2002), 794–797

34. A.M. Namboodiri, A.K. Jain, Retrieval of on-line hand-drawn sketches, in *Proceedings of the 17th International Conference on Pattern Recognition (ICPR'04)* (2004), pp. 642–645
35. K. Santosh, C. Nattee, B. Lamiroy, Spatial similarity based stroke number and order free clustering, in *Proceedings of the 12th International Conference on Frontiers in Handwriting Recognition* (2010), pp. 652–657
36. S. Parui, A. Mittal, Similarity-invariant sketch-based image retrieval in large databases, *Computer Vision–ECCV 2014* (Springer, Cham, 2014), pp. 398–414

A Multilayered Model for Artificial Intelligence of Game Characters as Agent Architecture

Youichiro Miyake

Abstract As all mathematics have a beautiful structure, an inner mind model of Artificial Intelligence has a grand architecture. It consists of information flow and software modules. In this twenty years, an agent's inner intelligence model has been studied and developed by many game AI programmers in game titles. A whole image of a current agent's intelligent model is explained.

Keywords Agent architecture · Blackboard architecture · Information flow · Affordance · Umwelt

1 Grand Design of an Agent's Mind

A game character lives in a game world. In these twenty years, a game world becomes much larger and more complex drastically. AI of a game character (character's mind) is required to be more intelligent, so it also becomes to have a strong intelligent structure in its mind. It is called "Agent Architecture", which is originally developed in robotics.

An agent architecture is a fundamental architecture which connects a game world and an agent's mind and body (Fig. 1). It consists of some modules which have a specific function, such as a sensor module, knowledge-making (recognition) module, decision-making module, motion-making module, and memory module. For example, a sensor module gathers information of game world. A knowledge-making module forms knowledge about the game world by using the information which a sensor module delivers. A decision-making module make a decision by joining some of knowledge together. A motion-making module makes a motion by following the decision.

Y. Miyake (✉)
Digital Game Research Association Japan, 3-2-1-336, Kamiasao, Aso-ku, Kawasaki-shi, Kanagawa-ken 215-0021, Japan
e-mail: y.m.4160@gmail.com

Y. Dobashi and H. Ochiai (eds.), *Mathematical Progress in Expressive Image Synthesis III*, Mathematics for Industry 24,
DOI 10.1007/978-981-10-1076-7_7

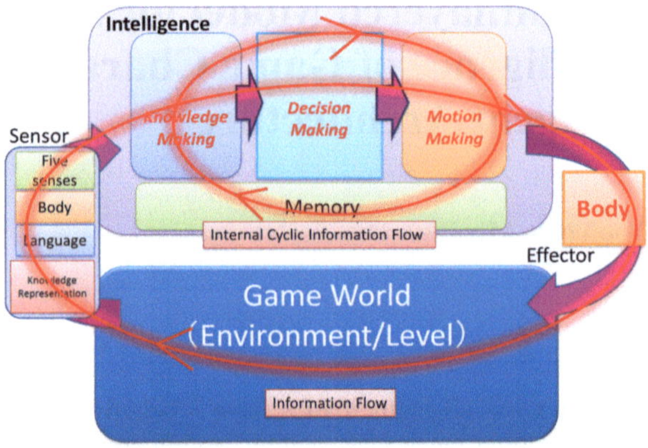

Fig. 1 Agent Architecture for a game agent [1]. It shows a relation of an agent and a world. It separates a world and agent via sensors and effectors. A *big circle* means an information flow through a game world and agent inside, which connects a world and an agent itself. A *small circle* is an information flow in agent architecture, which organizes data in agent architecture

2 Information Flow

An information flows through all modules and a game world as a process of an agent architecture and a world (Fig. 1). The flow is called "Information flow". An information flow connects an agent and a game world. The information keeps flowing while each module stores temporary information on memory module. So some blocks of information are stacked on memory module with specific format. These stacked memories are used by any module. An information flow has the information of the game world such as enemy's position, a terrain and some objects, and an event which happens in the game world. They consist of some symbols and float numbers. A form of data in information flow for an object, a terrain, or an event is called Knowledge Representation (KR). For example, an enemy's KR is a collection of the coordinate, the velocity vector, the enemy type (symbol), the weapon type (symbol), HP (numerical) and magic power (MP, numerical). An KR for an object is a structured data including affordance (afforded actions to the object), the position, and the object type. In a game, there are many enemies and objects, so an information flow has much information. They keep transformed in an agent architecture. In an agent architecture, information flow is abstracted to information flow in a higher level. Information are written on the higher blackboard. This abstraction process is iterated, so an information flow is repeatedly abstracted to higher level. By transforming data of information flow to more abstract data, AI creates the abstract recognition of a game world. The blackboard architecture becomes multi-layered system where some of KS combines each blackboard layer and form an abstraction process (Fig. 3).

3 Blackboard Architecture in Agent Architecture

A blackboard architecture is a simple architecture with one blackboard, some of KS (Knowledge Source), and an arbiter (Fig. 2, left image). The system is such as that some KSs write and read information on a blackboard, and the action tim-

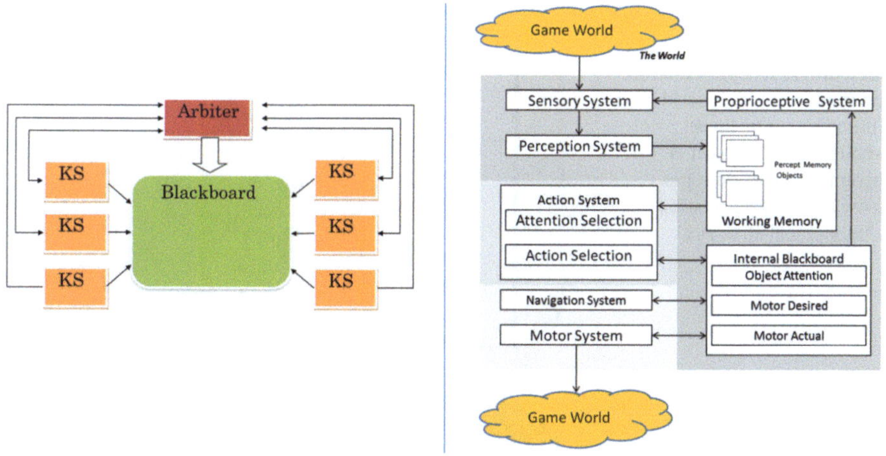

Fig. 2 The *left figure* is blackboard architecture with a blackboard, several KSs, and one arbiter [2]. The *right figure* is agent architecture applied blackboard architecture. It has 6 KS processing modules in the right part and two blackboards in the left part [3]

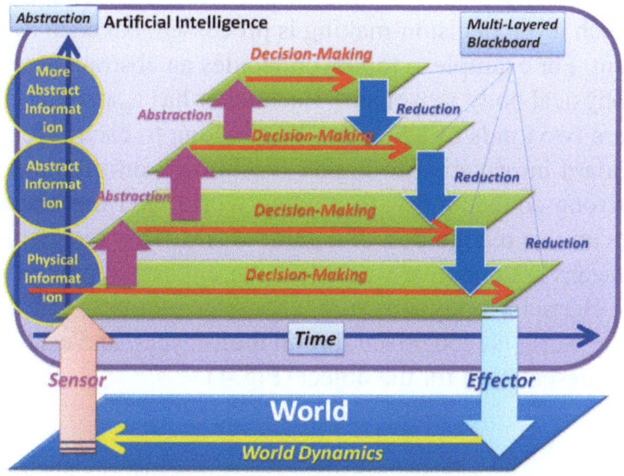

Fig. 3 A multilayered Blackboard architecture. It has four blackboards. The lowest blackboard represents a physical body, and it is strongly connected with a game world via sensors and effectors. There are three higher layers which have abstracted information of the lower layer. Each layer has a decision-making module. It has a different role and scale to the game world

ing is controlled by the arbiter. An agent architecture of a game character uses the blackboard architecture. The design is to separate all processing modules (such as knowledge-making, decision-making, and motion-making module) independently from memory module (Fig. 2, right image). A multi-layered blackboard architecture consists of some blackboards and KSs (Fig. 3). For example, four layered blackboard architecture is like this: In the lowest layer, detailed information of the game world is directly written on the lowest blackboard by KSs (sensor modules). In the second layer, more abstract information is written by abstracting detailed information of the lowest layer on the second by KS. In the third layer, more abstract information is written by abstracting the information of the second layer on the third blackboard by

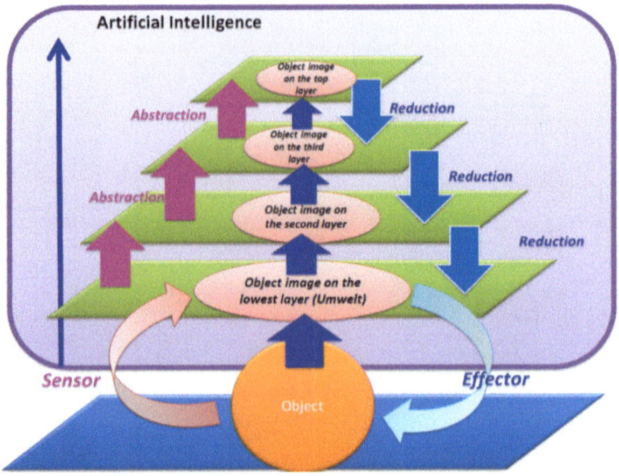

Fig. 4 An object representation. Each layer has a different representation for an object in the world. Through a multi-layer architecture, an object is recognized with multiple knowledge representations in a hierarchical structure

KS. The lowest layer expresses a physical body. This lowest layer is connected with a game world with a physical body and sensors. The information flow on the lowest layer is most fundamental flow connecting a character body and a game world. In each layer, decision-making is processed, but each role of decision-making is different. For example, a top layer decides an abstract goal while a lowest layer decides a physical body movement. Each layer has a specific role for decision-making. There are two kinds of KS; some KSs belong to each layer and some KSs abstract blackboard information to higher blackboard information. The lowest body layer has a strong connection with a game world, and the lowest blackboard forms an agent primitive recognition of a game world. This recognition is called "Umwelt", which means a subjective world [4]. An environmental world representation depends on a character body structure, character motions, and the life-style. And an object in the world is recognized throuh multilayers, and each layer forms a different abstract representation for the object (Fig. 4).

References

1. Y. Miyake, Current status of applying artificial intelligence for digital games. J. Jpn. Soc. Artif. Intell. 1 **30**, 45–64, Tokyo (2015), http://id.nii.ac.jp/1004/00000517/
2. D. Isla, B. Blumberg, Blackboard architectures. AI Game Program. Wisdom **1, 7.1**, 333–344 (2002). Boston
3. R. Burke, D. Isla, M. Downie, Y. Ivanov, B. Blumberg, Creature smarts: the art and architecture of a virtual brain. GDC 2001, http://naimadgames.com/publications.html
4. J. Uexküll, *Streifzüge durch die Umwelten von Tieren und Menschen: Ein Bilderbuch unsichtbarer Welten* (Rowohlt, Reinbek, 1956)

Visual Media Culture Supported by Illusion of Depth

Kokichi Sugihara

Abstract There is an enormous difference between the real world and its images; the real world has three dimensions, whereas images have only two. In spite of this difference, we can enjoy visual media without the need to put forth any special effort. Why can we do this? This question can be partly answered by studying the illusion of depth. It seems that human brains try to recover the depth from images with strong preference for special subclasses of objects, such as rectangular solids. This also suggests that visual media culture is fragile. We discuss this point using various depth illusions, such as impossible objects and impossible motions.

Keywords Depth perception · Impossible object · Impossible motion · Optical illusion · Visual media

1 Introduction

We use two eyes to observe objects in the real world, and this allows us to determine the distances to those objects. This is based on the mathematical fact that each point on an object can be located in three-dimensional space as the point of intersection of the two rays of sight from our eyes to the target point. This function is called binocular stereo vision [1, 3, 4].

However, when we observe images, such as photographs and movies, the visual data include information obtained through only one eye, because a camera has only a single lens center. Thus, there is no explicit information about the distance to an object. Nevertheless, we usually perceive a distance. This function is sometimes called monocular stereo vision [1, 7].

Of course, we use two eyes to observe an image, but this will just give us information about the distance to the image, not to the objects contained in the image.

K. Sugihara (✉)
Meiji University/JST, CREST, 4-21-1 Nakano, Nakano-ku,
Tokyo 164-8525, Japan
e-mail: kokichis@meiji.ac.jp

Y. Dobashi and H. Ochiai (eds.), *Mathematical Progress in Expressive Image Synthesis III*, Mathematics for Industry 24,
DOI 10.1007/978-981-10-1076-7_8

Nowadays, we are surrounded by various visual media, and we are apt to forget the differences between seeing objects directly and seeing them through images. However, there are enormous differences between binocular stereo vision and monocular stereo vision. The former is based on a mathematical principle, while the latter is not.

Monocular stereo vision is based on guessing the distances, and consequently, the results are very fragile. We will investigate this fragility by considering the optical illusions induced by impossible objects, impossible motions, and ambiguous cylinders.

2 Degrees of Freedom in Object Reconstruction

If we are given an object, a viewpoint E, and an image plane I, the image of the object projected on the image plane with respect to the center of projection at E is uniquely determined. However, even if we are given E, I, and the image, the original object is not unique. There is freedom in reconstructing the object from the image. This freedom can be mathematically specified in the following way [6].

We assume that the object is a polyhedron, that is, a solid bounded by planar faces. Suppose that the viewpoint E is at the origin of an (x, y, z) coordinate system, and that the image is fixed on the plane $z = 1$. Assume that the object in the image contains n vertices and m faces. Let $v_i' = (x_i, y_i, 1)$ be the ith vertex on the image. The original counterpart v_i of this vertex should be on the ray of sight emanating at the origin and passing through v_i'. Hence, the original vertex can be expressed by $v_i = (x_i/t_i, y_i/t_i, 1/t_i)$, where t_i is an unknown variable.

Let f_j be the plane containing the jth face of the object, and let

$$a_j x + b_j y + c_j z + 1 = 0 \tag{1}$$

be the equation for f_j, where a_j, b_j, and c_j are unknown variables.

Suppose that the vertex v_i is on the face f_j. Then, we can substitute the coordinates of v_i into Eq. (1), and we get

$$a_j x_i + b_j y_i + c_j + t_i = 0. \tag{2}$$

This is linear in the unknowns t_i, a_j, b_j, and c_j.

For each pair of a vertex and the face in which it is contained, we have an equation similar to Eq. (2), and hence we have a system of linear equations, which we denote by

$$Aw = 0, \tag{3}$$

where $w = {}^{t}(t_1, \ldots, t_n, a_1, b_1, c_1, \ldots, a_m, b_m, c_m)$ is a vector of unknown variables, and A is a constant matrix.

This system of equations contains $n + 3m$ unknown variables, and hence there are

$$n + 3m - \text{rank}(A) \tag{4}$$

degrees of freedom when interpreting an object from the image.

3 Depth Illusions

For a thick object, Eq. (4) specifies at least four degrees of freedom. This freedom enables us to design unfamiliar objects that will appear to be familiar ones; that is, this freedom enables us to create various optical illusions [6]. We will discuss some of these illusions below.

3.1 Impossible Objects

There is a class of pictures called "pictures of impossible objects" [5, 14]. This class of pictures gives us the impression that we are observing a three-dimensional structure, but at the same time, that structure is impossible. An famous example of this is the endless loop of stairs in the print "Ascending and Descending" (1960) [2] by the Dutch artist M.C. Escher; a simplified drawing of this staircase is shown in Fig. 1a.

The impression of impossibility comes from our knowledge that stairs consist of horizontal and vertical plates. However, the freedom specified in Eq. (4) suggests that there are other objects that can produce the same image, and we can construct a three-dimensional solid whose projection coincides with the drawing. Figure 1b shows an example of a solid that looks the same as Fig. 1a, c is another view of the same solid.

Another example is given in Fig. 2, where (a) shows a drawing of an impossible object, (b) shows a solid, and (c) shows another view of that solid.

These examples show that the impossible objects portrayed in pictures are not necessarily impossible; some of them can be realized as actual three-dimensional solids. This phenomenon shows that the human brain cannot always correctly extract solids from images.

3.2 Impossible Motions

The freedom in the choice of objects represented in images also enables us to design solids that appear to be ordinary but motions inserted to the solids seem to be physically impossible [10].

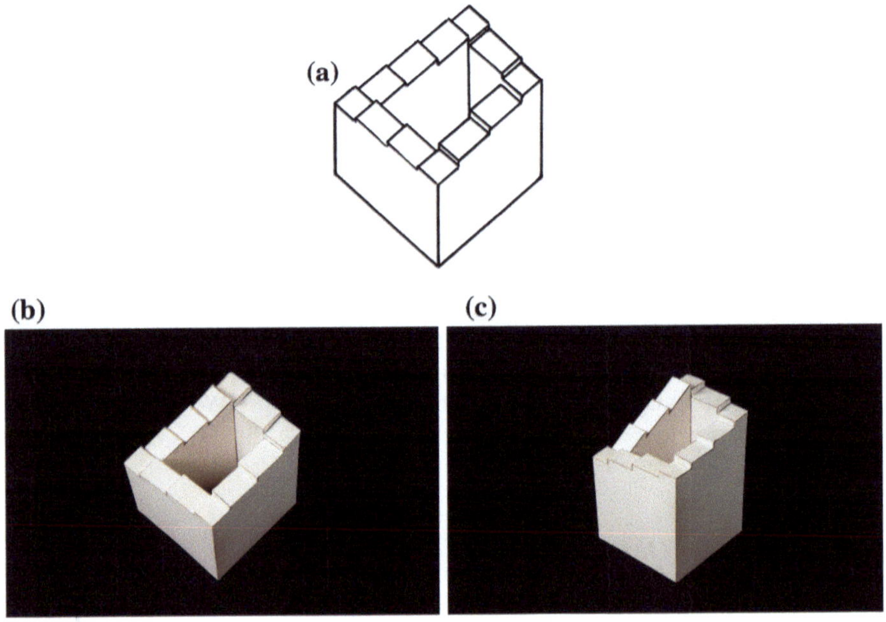

Fig. 1 Impossible object "Endless Loop of Stairs": **a** drawing; **b** a three-dimensional realization of the drawing shown in **a**; **c** another view of the solid shown in **b**

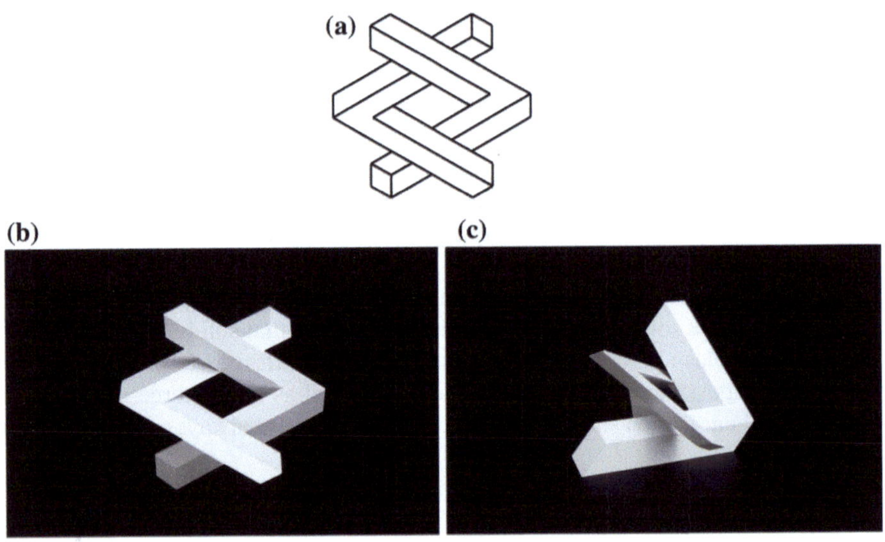

Fig. 2 Impossible object "Two L's": **a** drawing; **b** a three-dimensional realization of the picture shown in **a**; **c** another view of the solid shown in **b**

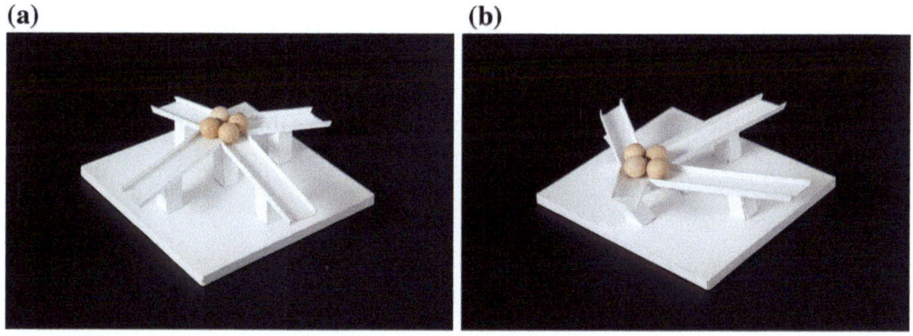

Fig. 3 Impossible motion "Magnet-Like Slopes": **a** view from which balls appear to be rolling uphill; **b** another view of the same solid

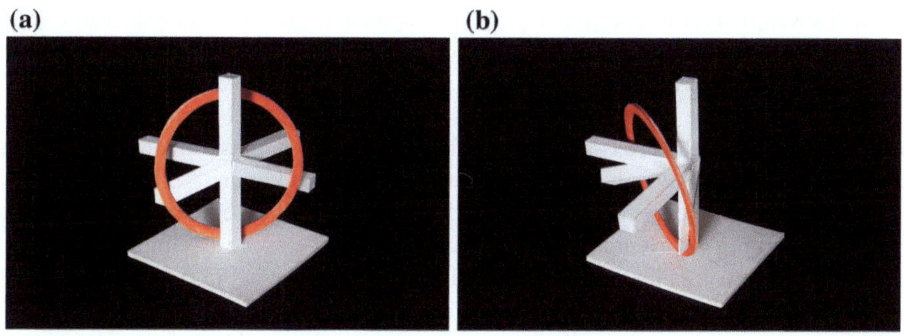

Fig. 4 Impossible motion "Four Perches and a Ring": **a** a ring hangs from the perches in a strange way; **b** another view of the same solid

An example is shown in Fig. 3. In Fig. 3a, we see what appears to be a solid that consists of four sloped ramps, each sloping downward in a different direction from the center, which appears to be the highest point. However, if balls are placed on any of the ramps, they will roll toward the center, as if they are defying the law of gravity. Actually, the center is the lowest point, as shown in Fig. 3b, and the balls simply roll downhill, as expected [8].

Another example is shown in Fig. 4. In Fig. 4a, the solid appears to consist of a pole and four horizontal perches that cross at right angles. However, a flat ring hangs in such a way that it passes behind the pole but in front of all four perches. The actual shape of the solid is shown in Fig. 4b; the four perches are all behind the pole [9].

A solid that generates an impossible motion illusion can be constructed in the following way. First, we generate a description of an ordinary three-dimensional object; this can be done using a solid modeling system. Next, we select a viewpoint and an image plane, and we generate the projected picture of the object; this can be done using computer graphics techniques. Finally, we generate and solve the associated system of Eq. (3). Due to the degrees of freedom specified by Eq. (4), there are infinitely many solutions, among which we can choose an object that can create a motion that appears to be impossible.

One interesting observation is that the impossible motion is perceived even if the observer is told the true shape of the object. This implies that the impossible motion illusion cannot be removed, even if we know the truth.

3.3 *Ambiguous Cylinders*

The third class of depth illusion is a class of ambiguous cylinders, whose appearance changes drastically when they are reflected in a mirror [11–13]. An example is shown in Fig. 5a. The roof of the garage appears to be round when it is seen directly, but when viewed in a mirror, it appears to be corrugated. Figure 5b shows the same solid seen from another angle. From this image, we see that the true shape of the roof is neither of the appearances seen in Fig. 5a.

Another example is given in Fig. 6. As seen in Fig. 6a, the object appears to be a cylinder for which the cross-section is a flower shape, while its mirror image shows a butterfly shape. Figure 6b is another view of the same cylinder.

Both of these objects can be produced by sweeping a line segment through space without changing its orientation. Let us call this type of surface a cylindrical surface; the solid in Fig. 6 is a closed cylinder, while the garage roof in Fig. 5 is an open cylinder. The length of the cylinder measured along the axis direction (i.e., the direction of the sweeping line segment) is the same at every point on the surface. The human brains seem to interpret the edge of the cylinder as the intersection of the cylinder and a plane perpendicular to the axis of the cylinder. This causes the ambiguous cylinder illusion.

Ambiguous cylinders can be created because of the freedom in the reconstruction of objects. This freedom enables us to design solids whose projections onto two

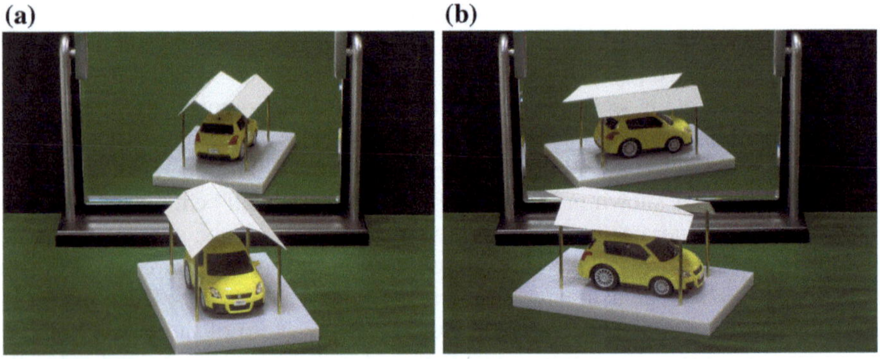

Fig. 5 Ambiguous cylinder "Ambiguous Garage Roof": **a** a solid and its mirror image seen from a special viewpoint; **b** the same solid seen from a general viewpoint

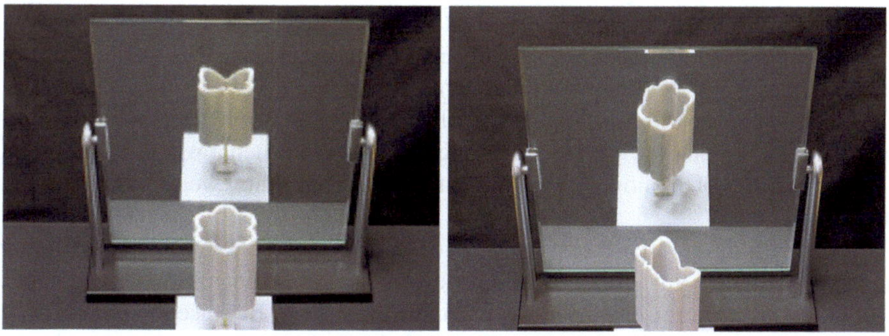

Fig. 6 Ambiguous cylinder "Flower and Butterfly": **a** a solid and its mirror image, as seen from a special viewpoint; **b** the same solid seen from a general viewpoint

Fig. 7 Computation of a space curve that realizes two given planar curves

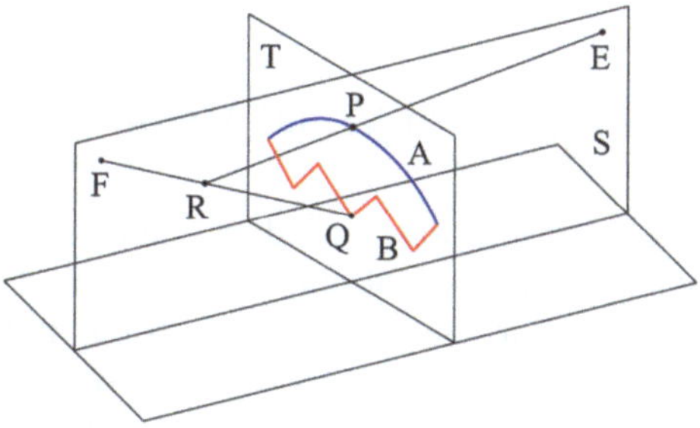

prespecified directions give desired pair of two-dimensional shapes, as explained below.

Let $A(s)$ and $B(s)$ be two planar curves, where s is a parameter such that $0 \le s \le 1$. As shown in Fig. 7, we assume that the two curves are embedded in a vertical plane T, they are monotone in the horizontal direction, and the left and right terminal points coincide, that is, $A(0) = B(0)$ and $A(1) = B(1)$.

Let E and F be two points that are not included in T. We want to construct a space curve $C(s)$ that appears to be $A(s)$ when it is seen from viewpoint E and that appears to be $B(s)$ when it is seen from viewpoint F. For this purpose, we choose sufficiently many values of $s \in [0, 1]$, and for each $P = A(s)$, we do the following.

Step 1. Construct plane S that contains the three points E, F, and P.
Step 2. Find point Q, which is at the intersection of the curve B with the plane S.
Step 3. Find point R on the intersection of the line EP with the line FQ, and set $C(s) = R$.

In practice, we choose a finite number of points $s_1 = 0 < s_2 < \cdots < s_{n-1} < s_n = 1$, compute the values $C(s_i)$, and then connect them to get a polyline $(C(s_1), C(s_2), \ldots, C(s_n))$ that approximates the space curve $C(s)$.

Note that the point Q in Step 2 can always be found, and it will be unique, because $B(s)$ is monotone in the horizontal direction, $A(0) = B(0)$, and $A(1) = B(1)$. Furthermore, the point R in Step 3 can always be found, and it will be unique, because the two lines EP and FQ are both on the plane S.

Next, we choose a line segment, say L, that is perpendicular to the plane S. Without changing its orientation, we move the line segment L in such a way that one of the terminal points of L moves along the curve $C(s)$, and thus we obtain the swept surface.

The resulting surface is a template for constructing the ambiguous cylinder. Indeed, if we view this surface from E, the cross-section coincides with $A(s)$, and if we view it from F, the cross-section coincides with $B(s)$. Because the human brain is apt to interpret this as the intersection of the cylinder with a plane perpendicular to the axis of the cylinder, we can expect that humans will perceive $A(s)$ from E and $B(s)$ from F.

4 Fragility of Visual Media

The depth illusions considered here suggest that it is difficult to interpret images as three-dimensional objects. However, it seems that the visual media in our daily lives, such as photographs, television, movies, and other videos, are used with the assumption that the users will perceive the correct depths. Therefore, there is a gap between this assumption and human abilities.

A primary use of visual media is to portray reality, but we can intentionally create visual media that result in incorrect impressions of the real world.

One common example of this is the photographs used in advertisements for real estate, which frequently give the impression that the rooms in a house are much larger than their actual size. We might say this is also an illusion. However, it portrays something that does not differ from reality except in the perceived size.

On the other hand, the optical illusions presented in Sect. 3 are more extreme, because the perceived shapes of the objects are very different from the actual shapes of the real objects. For example, ramps are perceived to slope in the opposite direction in the "Magnet-Like Slopes" illusion, and with ambiguous cylinders, the appearance is completely changed in the mirror image. These depth illusions suggest that visual media, such as photographs and movies, run serious risks of miscommunicating the shapes of objects. In particular, this implies the following two risks.

First, we are apt to think that there is little difference between seeing objects directly and viewing images of them. However, there is an important difference: in the former, we use binocular stereo vision, while in the latter, we use monocular stereo vision. Therefore, shape information may be distorted by visual media, even if this is not intentional.

Second and more seriously, visual media can be used to intentionally distort shape data, as we have seem in the various depth illusions in Sect. 3. We can intentionally cause viewers to perceive the orientation of a slope to be the opposite direction from the actual slope, as we saw in the impossible motion illusion. We can intentionally cause two viewers who see an object from different viewpoints to have very different perceptions of it.

Creatures got pairs of eyes during the middle of Cambrian period, which was about 500 million years ago. Since then, our brains have had a long time to learn binocular stereo vision. On the other hand, visual media technology, such as photographs and movies, have appeared within the past few hundred years. If we change the scale so that pairs of eyes evolved one year ago, then visual media technology was developed only 30 s ago. Therefore, we have not yet had time to evolve monocular stereo vision. This then leads to the serious illusions in depth perception that can result from images. It is thus necessary to be aware of this aspect of visual media technology.

5 Concluding Remarks

We have shown various depth illusion phenomena, such as impossible objects, impossible motions, and ambiguous cylinders. They all suggest that the monocular stereo function in our brains is fragile and that it can sometimes make a serious mistake in perceiving the depth to a surface, and consequently, the shape of an object. This might be due to the rapid and recent development of visual media technology, which has occurred during a period that is too short for our brains to evolve monocular stereo vision.

It is important to recognize this shortcoming when using visual media technology, and to be careful to avoid errors in depth perception.

Acknowledgments This work was partly supported by the Grants-in-Aid for Basic Scientific Research No. 24360039 and for Challenging Exploratory Research No. 15K12067.

References

1. D.H. Ballard, C.M. Brown, *Computer Vision* (Prentice-Hall, Englewood Criffs, 1982)
2. M.C. Escher, *The Graphic Work* (Taschen GmbH, Koln, 2004)
3. R.L. Gregory, *The Intelligent Eye* (Werdenfield and Nicolson, London, 1970)
4. D. Marr, *Vision* (W. H. Freeman and Company, New York, 1982)
5. L.S. Penrose, R. Penrose, Impossible objects: a special type of visual illusion. Br. J. Psychol. **49**, 31–33 (1958)
6. K. Sugihara, *Machine Interpretation of Line Drawings* (MIT Press, Cambridge, 1986)
7. K. Sugihara, Three principles in stereo vision. Adv. Robot. **1**, 391–400 (1986)
8. K. Sugihara, Impossible Motion Magnet-Like Slopes, http://illusionoftheyear.com/cat/top-10-finalists/2010/
9. K. Sugihara, Impossible Motion 2. http://www.youtube.com/watch?v=1dISzq4IPc
10. K. Sugihara, Design of solids for antigravity motion illusion. Comput. Geom. Theory Appl. **47**, 675–682 (2014)
11. K. Sugihara: Design of ambiguous cylinders, *10th Asian Forum on Graphic Sciences*, Bangkok, August 4–7, 2015

12. K. Sugihara, *Joy of Ambiguous Solids: How to Make Anomalous Objects That Change Their Appearances in a Mirror* (Amazon Services International Inc, Kindle, 2015)
13. K. Sugihara, Height reversal generated by rotation around a vertical axis. J. Math. Psychol. **68–69**, 7–12 (2015)
14. J. Timothy Unruh, *Impossible Objects* (Sterling Publishing Co., Inc., New York, 2001)

Wang Tile Modeling of Wall Patterns

Alexandre Derouet-Jourdan, Yoshihiro Mizoguchi and Marc Salvati

Abstract Wall patterns are essential in the creation of textures for visually rich buildings. Particularly, irregular wall patterns give an organic and lively feeling to the building. In this chapter, we introduce a modeling method for wall patterns using Wang tiles which are known for creating aperiodic tiling of the plane under certain conditions. We introduce a class of Wang tiles and prove that any rectangle with border constraints and bigger than a 2×2 rectangle can be tiled. We use this proof to derive a tiling algorithm that is in linear time. Finally, we give some results of our algorithm and compare the computation time with previous Wang tiling algorithms introduced in computer graphics.

Keywords Texture synthesis · Wall patterns · Wang tiles · Tiling algorithms

1 Introduction

Designing interesting and visually rich buildings requires the creation of complex wall or ground textures. Particularly, it requires the creation of wall patterns. To give a lively feeling to its creation, the artist may need to add a stochastic aspect to the distribution of bricks, so that the wall appears less regular and more organic. An example of such a texture is given in Fig. 1.

A. Derouet-Jourdan (✉)
OLM Digital Inc./JST CREST, Dens Kono Bldg. 302, 1-8-8 Wakabayashi, Setagaya-ku, Tokyo 154-0023, Japan
e-mail: alexandre.derouet-jourdan@olm.co.jp

Y. Mizoguchi
Institute of Mathematics for Industry, Kyushu University/JST CREST, Fukuoka 819-0395, Japan
e-mail: ym@imi.kyushu-u.ac.jp

M. Salvati
OLM Digital, Inc., Mikami Bldg. 2F, 1-18-10 Wakabayashi, Setagaya-ku, Tokyo 154-0023, Japan
e-mail: marc.salvati@olm.co.jp

Y. Dobashi and H. Ochiai (eds.), *Mathematical Progress in Expressive Image Synthesis III*, Mathematics for Industry 24,
DOI 10.1007/978-981-10-1076-7_9

Fig. 1 Example of a wall texture created by an artist from OLM Digital, Inc.

The process of creating such patterns is tedious but can be achieved without a high level of human control. Basically, the artist proceeds by selecting the borders constraints, then fill the inside of the wall by adding rectangles and correcting the distribution until satisfaction. It is a long process that cannot be used to create very large texture. In this situation, in order to texture large building, the artist creates an auto-tileable pattern that is repeated on the building, introducing a visual periodicity.

In this chapter, we propose to replace this tedious process by an algorithm that designs wall patterns automatically. To this intent, we model the bricks intersections with Wang tiles as a specific class of Wang tiles. Accounting for border constraints is important to create repeatable patterns or to stitch together different patterns. We prove that any rectangle with border constraints and bigger than a 2×2 square can be tiled with our Wang tiles as soon as the number of colors in the Wang tiles is greater than 2. This proof is inductive and provides a tiling algorithm that is fast (linear time) and accounts for border constraints.

2 Related Works

Tiling a rectangle (or a plane) means filling it with elementary shapes, the tiles, such that the interior of tiles do not overlap and every point of the rectangle belongs to at least one tile. Tiling is a very powerful tool to create visually rich 2d structures. For more details on tiling and tiling in computer graphics, please consult [5].

2.1 Wang Tiles

In this chapter, we are interested in a special class of tiles, Wang tiles [9]. A Wang tile is a unit square with colored edges. Tiling with Wang tiles requires that the tiles are placed on a regular grid, edges to edges. Tiles from the tile set can be used as

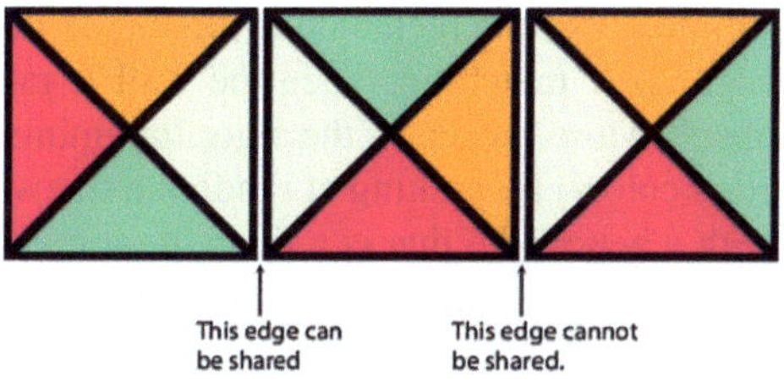

Fig. 2 Examples of Wang tiles. The *middle tile* can share an edge with the *left tile* because the colors match. It can not share an edge with the *right one* because the colors don't match

many times as required, but two tiles can share an edge only when they have the same color on the shared edge. We give three examples of Wang tiles in Fig. 2.

The Wang tiles have the interesting property of creating irregular patterns. Although Wang conjectured that if a finite Wang tile set were tiling the plane, then it was possible to tile the plane with the same tile set but periodically, this conjecture was disproved by Robert Berger in [1] when he introduced a Wang tile set that could only tile the plane aperiodically. Since then, several other aperiodic tile sets have been introduced [3, 6]. In computer graphics, Wang tiles have been used to create non periodic textures or points distributions automatically [2, 7].

2.2 *Tiling Algorithms*

In the computer graphics community, in the context of texture synthesis, two tiling algorithms have been elaborated to tile a rectangle with a given set of Wang tiles.

2.2.1 Sequential Tiling Algorithm

The easiest approach to tile a rectangle with a set of Wang tiles is to tile sequentially, row by row (or column by column) the rectangle, choosing at random a tile that matches the tiles already inserted in the rectangle. This algorithm was introduced in [2] for textures and point distribution generation. It has one important requirement on the tiles, that is every color-combination for left edge and top edge must exist in the tiles.

Advantages of this algorithm is to be easy to implement and fast, since it operates in linear time. One major drawback is that it can be hard to satisfy boundary constraints with this algorithm, as they introduce strong requirements on the tile set—that our tile sets do not satisfy.

2.2.2 Stochastic Tiling Algorithm

Another tiling algorithm for Wang tiles, the stochastic tiling algorithm, has been introduced in [7] to account for border constraints. The idea of the algorithm is very simple. First, the rectangle is filled with random Wang tiles, without accounting for edge color matching. Then the wall is randomly "corrected" until all edge colors match. More precisely, the algorithm minimizes the number of errors (non matching edge colors) by picking at random a tile with non matching edges, then replacing it with a better tile, that is a tile with at most the same number of errors. By iterating this process, the algorithm converges to a distribution of tiles with all edge colors matching.

One big advantage of this algorithm is that it produces a tiling satisfying the border constraints. Although in practice this algorithm works well, it is not proven that it will always converge. It is also difficult to estimate the number of iterations until convergence. In a production context, it is very important to give some guarantee of convergence and some estimation of computation time.

3 Wang Tiles for Walls

In our approach, the Wang tiles are containing the corners of the bricks in our wall. In other terms, the Wang tiles are all the way 4 bricks can connect in the wall. We suppose that the edges of the bricks are axis-aligned and we add the constraints that

- for aesthetics reasons, the 4 bricks do not share a vertex—they do not make a cross,
- each tile is traversed with a straight line, either vertically or horizontally.

The Wang colors are then the positions where the bricks intersect the tile edge, see Fig. 3 for an example. For instance, in the results presented in Sect. 4, we consider a color set C containing five Wang edge colors, corresponding to bricks intersecting the tile edge at lengths 0.25, 0.45, 0.5, 0.55, 0.75.

Such a tile set has a very interesting property. It is indeed easy to show that given a surrounding with two or three tiles, it is possible to find a tile in the tile set that has no erroneous edges shared with this surrounding. In other words, if we fix up to three edges colors, we can always find at least one tile in the tile set that matches those colors. This property does not hold for 4 edge colors constraints because for instance,

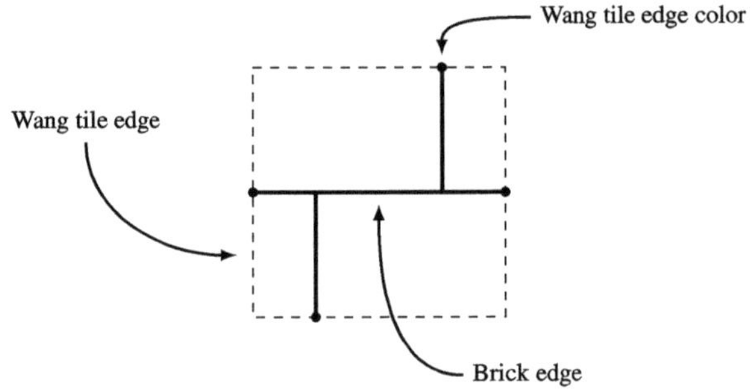

Fig. 3 We use Wang tiles to model the brick corners. The Wang tiles colors, that is the colors of the edges of the Wang tiles, represent the position of the intersection of the bricks edges with the edges of the tile

a tile with all its edges colors equal does not exist because it would introduce a cross in the tiles.

3.1 Definitions and Notations

Let C be a finite set of colors. A **Wang tile** is a function $w : \{t, l, b, r\} \rightarrow C$. Values $w(t)$, $w(l)$, $w(b)$, and $w(r)$ denote colors of top, left, bottom and right of a tile w respectively, as illustrated in Fig. 4. A Wang tile w is said in the class $\mathscr{W}$, the class of brick corners Wang tiles, if $((w(t) \neq w(b)) \wedge (w(l) = w(r)))$ or $((w(t) = w(b)) \wedge (w(l) \neq w(r)))$.

We define the set $W = \{w \mid w \in \mathscr{W}\}$ of all Wang tiles in class $\mathscr{W}$. Let P_{nm} be a rectangle with size $n \times m$, that is $P_{nm} = \{(i,j) \in \mathbf{N} \times \mathbf{N} \mid 1 \leq i \leq n,\ 1 \leq j \leq m\}$. The set of edges is $E = \{e_{i,j} \mid 1 \leq i \leq n,\ 0 \leq j \leq m\} \cup \{e'_{i,j} \mid 0 \leq i \leq n,\ 1 \leq j \leq m\}$.

A **boundary** coloring is a function $B_{nm} : \{e_{i0}, e_{im} \in E \mid 1 \leq i \leq n\} \cup \{e'_{0j}, e'_{nj} \in E \mid 1 \leq j \leq m\} \rightarrow C$. A **tiling** is a function $\mathscr{T} : P_{nm} \rightarrow W$ and we denote $\mathscr{T}(i,j)$ by $\mathscr{T}_{i,j}$. Given a tiling $\mathscr{T}$ and a tile $w \in W$, we denote $\mathscr{T}_{i,j \leftarrow w}$ the tiling obtained from $\mathscr{T}$ by replacing $\mathscr{T}_{i,j}$ with w. We call $S(\mathscr{T}_{i,j}) = \{e_{i,j-1}, e_{i,j}, e'_{i-1,j}, e'_{i,j}\}$ **surrounding** edges of a tiling $\mathscr{T}$ at a position (i,j), see Fig. 5. A **validation** map $\mathbf{c} : E \rightarrow \{\mathbf{true}, \mathbf{false}\}$ is defined by

$$\mathbf{c}(e_{i,j}) = \begin{cases} \mathbf{true} & (\mathscr{T}_{i,1}(t) = B(e_{i0}) \wedge (j = 0)) \text{ or} \\ & (\mathscr{T}_{i,j}(b) = \mathscr{T}_{i,j+1}(t) \wedge (1 \leq j \leq m-1)) \text{ or} \\ & (\mathscr{T}_{i,m}(b) = B(e_{in}) \wedge (j = m)), \\ \mathbf{false} & \text{otherwise,} \end{cases}$$

$$\mathbf{c}(e'_{i,j}) = \begin{cases} \mathbf{true} & (\mathscr{T}_{1,j}(l) = B(e'_{0j}) \wedge (i = 0)) \text{ or} \\ & (\mathscr{T}_{i,j}(r) = \mathscr{T}_{i+1,j}(l) \wedge (1 \leq i \leq n-1)) \text{ or} \\ & (\mathscr{T}_{n,j}(r) = B(e'_{nj}) \wedge (i = n)), \\ \mathbf{false} & \text{otherwise.} \end{cases}$$

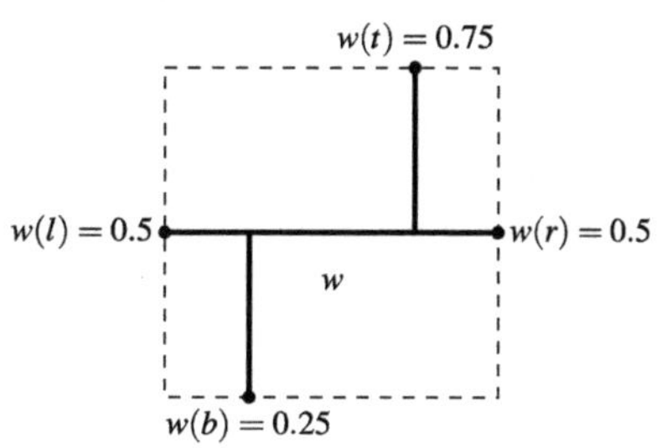

Fig. 4 Notations on one Wang tile as defined in Sect. 3.1

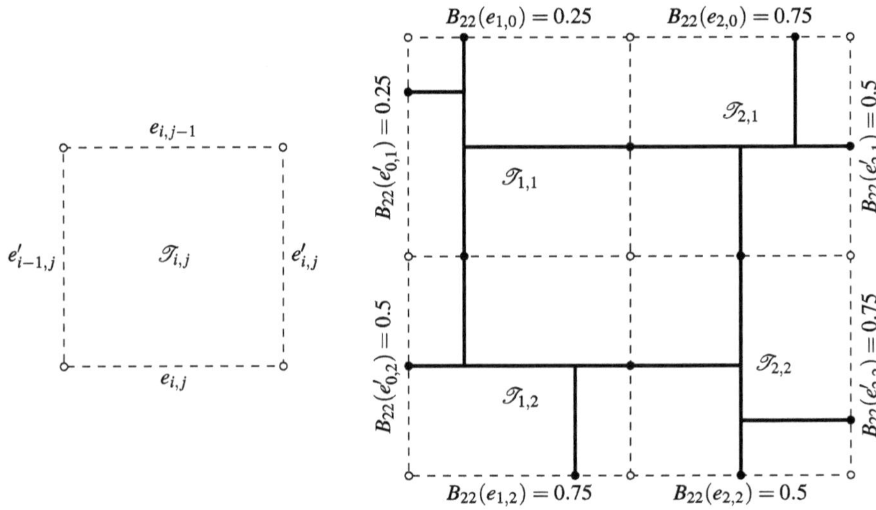

Fig. 5 Notations for a tiling $\mathscr{T}$ and its edges

A tiling $\mathscr{T}$ is **valid** if $\mathbf{c}(e_{i,j}) =$ **true** and $\mathbf{c}(e'_{i,j}) =$ **true** for all for edges $e_{i,j}$ and $e'_{i,j}$ in E. We call an edge e **erroneous** if $\mathbf{c}(e) =$ **false**. We denote $\mathbf{e}(\mathscr{T}_{i,j})$ the number of erroneous edges in $S(\mathscr{T}_{i,j})$. The total number of erroneous edges for a tiling $\mathscr{T}$ is denoted by $\mathbf{E}_{\mathscr{T}}$.

Definition 1 (*Tileable*) Let n and m be natural numbers. A rectangle P_{nm} is **tileable** if there exists a valid tiling $\mathscr{T}$ for any boundary coloring B_{nm}.

Definition 2 Let n and m be natural numbers and $n, m \geq 2$. A boundary coloring B_{nm} is a **torus** boundary coloring if $B_{nm}(e'_{0j}) = B_{nm}(e'_{nj})$ and $B_{nm}(e_{i0}) = B_{nm}(e_{im})$ for any $1 \leq i \leq n$ and $1 \leq j \leq m$. A valid tiling $\mathscr{T}$ for a torus boundary is called a **periodic tiling**.

3.2 Satisfiability of Border Constraints

The stochastic tiling algorithm accounts for border constraints but requires an important computation time. The sequential tiling algorithm is faster but does not account for border constraints. In this section, we prove that the border constraints can always be satisfied with our Wang tiles when the rectangle contains a 2×2 square. The proof induces a new algorithm, similar to the sequential tiling algorithm and that accounts for border constraints.

Proposition 1 *Let C be a color set.*

1. *P_{12} is not tileable (i.e. there exist boundary colorings that can not be satisfied).*

2. *P_{22} is not tileable if $|C| = 2$ (we show later that P_{22} is tileable when $|C| > 2$).*

Proof Let $c_1, c_2 \in C$ and $c_1 \neq c_2$.

1. We show an example of a boundary coloring B_{12} for which there is no valid tiling $\mathcal{T}$. If $B_{12}(e'_{01}) = c_1$, $B_{12}(e'_{02}) = B_{12}(e'_{11}) = B_{12}(e'_{12}) = B_{12}(e_{10}) = B_{12}(e_{12}) = c_2$ then there is no valid tiling. If $\mathcal{T}$ is a valid tiling, then $\mathcal{T}_{11}(l) = c_1$ and $\mathcal{T}_{11}(t) = \mathcal{T}_{11}(r) = \mathcal{T}_{12}(l) = \mathcal{T}_{12}(b) = \mathcal{T}_{12}(r) = c_2$. Since $\mathcal{T}_{11}(b) = c_2$ and $\mathcal{T}_{12}(t) \neq c_2$, we have $\mathcal{T}$ is not a valid tiling.
2. We define a boundary coloring B_{22} by $B_{22}(e_{10}) = B_{22}(e'_{01}) = B_{22}(e_{12}) = B_{22}(e_{22}) = B_{22}(e_{20}) = c_1$ and $B_{22}(e'_{02}) = B_{22}(e'_{21}) = B_{22}(e'_{22}) = c_2$. It is easy to check there is no valid tiling $\mathcal{T}$ if $|C| = 2$.

Lemma 1 *Let C be a color set, B a boundary coloring of P_{12}.*

1. *Let $B(e_{10}) = B(e_{12})$. There exists a valid tiling $\mathcal{T}$ if and only if*

$$(B(e'_{01}) = B(e'_{11}) \iff B(e'_{02}) = B(e'_{12})). \tag{1}$$

2. *Let $B(e_{10}) \neq B(e_{12})$. There exists a valid tiling $\mathcal{T}$ if and only if*

$$(B(e'_{01}) = B(e'_{11}) \vee B(e'_{02}) = B(e'_{12})). \tag{2}$$

Proof 1. Assume $B(e'_{01}) = B(e'_{11})$. If Eq. (1) holds then $B(e'_{02}) = B(e'_{12})$ and P_{12} admits a valid tiling. That is we choose tiles $\mathcal{T}_{11}$ and $\mathcal{T}_{12}$ with $\mathcal{T}_{11}(l) = \mathcal{T}_{11}(r) = B(e'_{01})$ and $\mathcal{T}_{12}(l) = \mathcal{T}_{12}(r) = B(e'_{02})$. We note that $\mathcal{T}_{11}(b) = \mathcal{T}_{12}(t) \neq B(e_{10})$ and $\mathcal{T}_{11}(b) = \mathcal{T}_{12}(t) \neq B(e_{12})$. Conversely, if P_{12} admits a valid tiling then $\mathcal{T}_{11}(l) = \mathcal{T}_{11}(r)$, $\mathcal{T}_{11}(t) \neq \mathcal{T}_{11}(b)$, $\mathcal{T}_{11}(t) = B(e_{10})$, $\mathcal{T}_{12}(b) = B(e_{12})$ and $\mathcal{T}_{12}(t) \neq \mathcal{T}_{12}(b)$. So we have $B(e'_{02}) = \mathcal{T}_{12}(l) = \mathcal{T}_{12}(r) = B(e'_{12})$. That is Eq. (1) holds.
Assume $B(e'_{01}) \neq B(e'_{11})$. The Eq. (1) means $B(e'_{02}) \neq B(e'_{12})$. It is easy to show that if $B(e'_{02}) \neq B(e'_{12})$ then there exists a valid tiling. And if $B(e'_{02}) = B(e'_{12})$ then there is no valid tiling.
2. Assume that Eq. (2) holds. It is easy to show there exists a valid tiling. Conversely, assume there exists a valid tiling. Since $B(e_{10}) \neq B(e_{12})$ and $\mathcal{T}_{11}(b) = \mathcal{T}_{12}(t)$, we have $\mathcal{T}_{11}(t) \neq \mathcal{T}_{11}(b)$ or $\mathcal{T}_{12}(t) \neq \mathcal{T}_{12}(b)$. That is $B(e'_{01}) = \mathcal{T}_{11}(l) = \mathcal{T}_{11}(r) = B(e'_{11})$ or $B(e'_{11}) = \mathcal{T}_{12}(l) = \mathcal{T}_{12}(r) = B(e'_{12})$.

Proposition 2 *Let C be a color set and ($|C| \geq 3$). A rectangle P_{nm} with a border coloring B_{nm} is tileable for all natural numbers n, m ($n, m \geq 2$).*

Proof **(i)** P_{22} is tileable.
We divide P_{22} into two P_{12} and the tileability is proved by matching a given boundary coloring to conditions of Lemma 1.
(ii) If P_{nm} is tileable then $P_{n(m+1)}$ is tileable.
Let $B_{n(m+1)}$ be a given boundary coloring for $P_{n(m+1)}$. We define a tiling $\mathcal{T}$ of $P_{n(m+1)}$ as follows. We can define $\mathcal{T}_{i(m+1)}$ $(1 \leq i \leq n-1)$ satisfying $\mathcal{T}_{1(m+1)}(l) =$

$B_{n(m+1)}(e'_{0(m+1)})$, $\mathscr{T}_{(i+1)(m+1)}(l) = \mathscr{T}_{i(m+1)}(r)$ and $\mathscr{T}_{i(m+1)}(b) = B_{n(m+1)}(e_{i(m+1)})$. Since each tile $\mathscr{T}_{i(m+1)}$ has only two conditions, we can choose a Wang tile $\mathscr{T}_{i(m+1)}$ satisfying conditions easily. Next we define $\mathscr{T}_{n(m+1)}$ satisfying the following three conditions $\mathscr{T}_{n(m+1)}(l) = \mathscr{T}_{(n-1)(m+1)}(r)$, $\mathscr{T}_{n(m+1)}(b) = B_{n(m+1)}(e_{n(m+1)})$ and $\mathscr{T}_{n(m+1)}(r) = B_{n(m+1)}(e'_{n(m+1)})$. Finally, we reduce the tiling problem of $P_{n(m+1)}$ to the problem of P_{nm}. Define a boundary coloring B_{nm} for the problem of P_{nm} by $B_{nm}(e_{i0}) = B_{n(m+1)}(e_{i0}), B_{nm}(e_{im}) = \mathscr{T}_{i(m+1)}(t), B_{nm}(e'_{0j}) = B_{n(m+1)}(e'_{0j})$ and $B_{nm}(e'_{nj}) = B_{n(m+1)}(e'_{nj})$, $(1 \leq i \leq n, 1 \leq j \leq m)$. Since we are assuming P_{nm} is tileable, we have a tiling $\mathscr{T}$ for P_{nm} with boundary coloring B_{nm} and finally we have a tiling $\mathscr{T}$ for $P_{n(m+1)}$ with boundary coloring $B_{n(m+1)}$.

(iii) If P_{nm} is tileable then $P_{(n+1)m}$ is tileable.

It is similarly proved with the result of (ii).

(iv) P_{nm} is tileable.

It is proved inductively using the results of (i), (ii) and (iii).

3.3 Boundary-Constrained Sequential Tiling Algorithm

The proof of Proposition 2 induces naturally a tiling algorithm for a rectangle P_{nm} with a constrained boundary coloring B_{nm}. This algorithm, called boundary-constrained sequential (BCS) tiling algorithm consists in picking at random a boundary coloring B_{nm} and then recursively tile the rectangle P_{nm} until we have $(n = m = 2)$. In practice, the recursivity is replaced by a sequential loop that visit the columns and then the rows to reduce the tiling to P_{22}. Then, the square P_{22} is tiled using Lemma 1. This tiling algorithm has a computation time linear with respect to nm, the total number of tiles in the considered rectangle, and accounts for boundary constraints. Particularly, the BCS tiling algorithm accounts for torus boundary condition, which is useful in practice to create auto-tileable patterns that allows to texture a mesh with a low memory footprint.

4 Results

In the following, we evaluate the performance of the new algorithm and compare it with the two algorithms from previous work. In this section, the Wang tiles are built with a color set C containing five Wang edge colors, corresponding to bricks intersecting the tile edge at lengths 0.25, 0.45, 0.5, 0.55, 0.75.

4.1 Performance

Figure 6, we display the result of running the combination of sequence and stochastic relaxation algorithm to create a tiling of 49×30 tiles. Using Wang tiles allows to

Fig. 6 Example of wall pattern created by the boundary-constrained sequential tiling algorithm. This tiling contains $49 \times 30 = 1470$ tiles and was computed in 1 ms. 1609 tiles were visited during this computation. The bricks colors are added in a post-process

create structures without visible repeated patterns. We added colors to the bricks in a post-process, for aesthetics reasons.

Table 1, we report the computation time of the three algorithm for the creation of a small size tiling (49×30) and a big size one (1000×1000). As we can see, the sequential algorithm is fast but does not create auto-tileable patterns since it can not account for border constraints. The stochastic algorithm accounts for the border constraints but is slower, as it visits more tiles during the computation (in this example, it visits 10 times more tiles). Using the combination of both algorithm allows to account for border constraints, while benefiting from the speed of the sequential algorithm, with only a limited overhead.

4.2 Limitations

Modeling the wall pattern structure using Wang tiles allows the creation of rich non-repeated patterns. The tiling algorithm are fast and it is possible to account for border constraints.

The next step to make it available in a production environment is to create shaders from those patterns, by adding a texture to the bricks and a proper gap between them. To this end, it is important to be able to use the tiling structure to have a direct control over the shapes of the bricks and the gaps between them. Moreover, artists often need control over the creative process. For instance, they may need to

Table 1 Results of the three algorithms for a rectangle of size 49 × 30 (middle column) and 2000 × 2000 (right column, given as an average on 20 runs)

Results	49 × 30			1000 × 1000	
Sequential	Visited tiles	Time	Tileable?	Visited tiles	Time
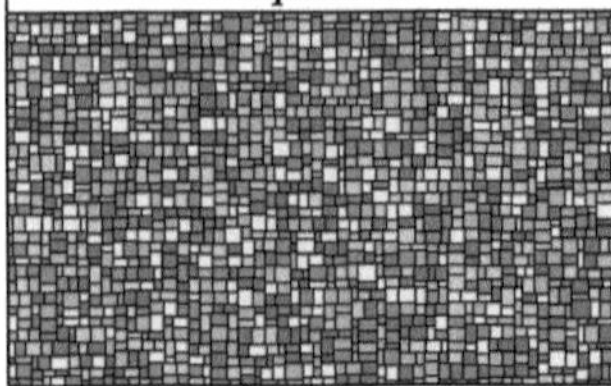	1470	1ms	No	1000000	716ms
Stochastic	Visited tiles	Time	Tileable?	Visited tiles	Time
	20324	30ms	Yes	13837844	41s
BCS	Visited tiles	Time	Tileable?	Visited tiles	Time
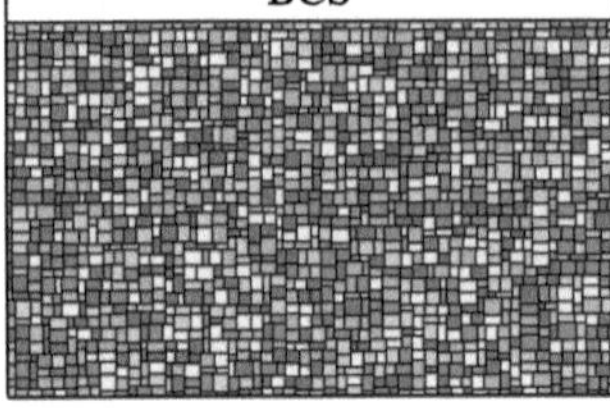	1470	1.4ms	Yes	1000000	1s

If the sequential tiling algorithm is fast, it does not produce a tileable pattern. The stochastic tiling algorithm produces such pattern but is slower. The boundary-constrained sequential tiling algorithm (BCS), with only a small overhead on the sequential tiling algorithm, create quickly tileable patterns

control the distribution of bricks, having bigger bricks in some region of the wall, or a regular pattern in another one. In terms of Wang tiles, it means that we need to add the possibility of using different Wang tiles for different region of the tiling. In that case, it becomes necessary to be able to create some transitions tiles between the tile sets to guarantee the tileability of the wall.

If the creation of wall patterns is automatic, it is still needed to create Wang tiles manually. It then becomes important to check that the new tiles can tile the rectangle and that the algorithms can be used to produce the tiling. In addition, while convergence of the stochastic algorithm is empirically verified, we do not have any proof of this claim. It is an important development as we need the stochastic algorithm to account for border constraints.

5 Conclusion

In this chapter, we introduced the modeling method for wall patterns using Wang tiles. Such a modeling approach allows us to use tiling algorithm to create efficiently wall patterns usable for texturing.

As future works, besides those we discussed above, we would be interested in creating 3d walls, by using 3d Wang tiles, Wang cubes [4]. We would then like to investigate how such a tool could be useful for volume rendering or 3d textures. In 2d, we proved that any rectangle with a boundary constraint is tileable if it is bigger than a 2×2 square. We believe that this property remains true for more complex polygon when they contain a 2×2 square. We would like to study this assumption and prove it, formally and using a theorem prover such as Coq [8].

Acknowledgments This work was supported by Core Research for Evolutional Science and Technology (CREST) Program "Mathematics for Computer Graphics" of Japan Science and Technology Agency (JST). The authors are grateful to the collaborators including Ken Anjyo, Ayumi Kimura, and Richard Roberts at OLM Digital, Hiroyuki Ochiai at Kyushu University and Yoshinori Dobashi at Hokkaido University for their valuable discussions.

References

1. R. Berger, *The Undecidability of the Domino Problem*, vol. 66 (American Mathematical Soc., Providence, 1966)
2. M.F. Cohen, J. Shade, S. Hiller, O. Deussen, Wang tiles for image and texture generation. ACM Trans. Graph. **22**(3), 287–294 (2003)
3. K. Culik, An aperiodic set of 13 Wang tiles. Discret. Math. **160**(1), 245–251 (1996)
4. K. Culik II, J. Kari, An aperiodic set of Wang cubes, in *J. UCS The Journal of Universal Computer Science* (Springer, 1996), pp. 675–686
5. C. Kaplan, *Introductory Tiling Theory for Computer Graphics* (Morgan & Claypool Publishers, San Rafael, 2009)
6. J. Kari, A small aperiodic set of Wang tiles. Discret. Math. **160**(1), 259–264 (1996)
7. J. Kopf, D. Cohen-or, O. Deussen, D. Lischinski, Recursive Wang tiles for real-time blue noise, in *Proceedings of the SIGGRAPH* (Citeseer, 2006)
8. The Coq development team. *The Coq proof assistant reference manual*. LogiCal Project (2004). Version 8.0
9. H. Wang, Proving theorems by pattern recognition-ii. Bell Syst. Tech. J. **40**(1), 1–41 (1961)

High-Resolution Visualization Library for Exascale Supercomputer

Yoshitaka Wada, Kohei Murotani, Masao Ogino, Hiroshi Kawai and Ryuji Shioya

Abstract This paper describes development of visualization library, which is named as LexADV_VSCG, for very large scale on the next generation supercomputer. At this moment, the next generation supercomputer, which is called as exa-scale computer, is not designed clearly. We predict the exa-scale computer and research the possibility of new software which is optimized for the exa-scale computer. In this work, we show how to visualize and deal with ultra large scale FE and particle based data with very fine resolution. LexADV_VSCG provide only simple API for drawing and rendering triangles and lines including transparent and solid colors. Since cross section generation is well required by engineers, additional advanced features, which are cross section generation and miscellaneous drawing functions, are provided by LexADV_VSCG.

Keywords Exa-scale computing · Parallel finite element method · Parallel particle-based method · Ultra high resolution

---

Y. Wada (✉)
Kindai University, 3-4-1 Kowakae, Higashiosaka, Osaka, Japan
e-mail: wada@mech.kindai.ac.jp

K. Murotani
University of Tokyo, 7-3-1 Hongo Bunkyo, Tokyo, Japan
e-mail: muro@sys.t.u-tokyo.ac.jp

M. Ogino
Nagoya University, Furo-cho, Chikusa, Nagoya, Japan
e-mail: masao.ogino@cc.nagoya-u.ac.jp

H. Kawai
Tokyo University of Science, Suwa, 5000-1 Toyohira, Chino, Nagano, Japan
e-mail: kawai@rs.tus.ac.jp

R. Shioya
Toyo University, 2100 Kujirai, Kawagoe, Saitama, Japan
e-mail: shioya@toyo.jp

Y. Dobashi and H. Ochiai (eds.), *Mathematical Progress in Expressive Image Synthesis III*, Mathematics for Industry 24,
DOI 10.1007/978-981-10-1076-7_10

1 Introduction

The large scale FE analysis would be bottleneck using supercomputer systems [1–4]. However a commodity computer system becomes popular and faster than ever and graphic accelerator becomes 20 times faster than 10 years before at moderate estimation. The Kei supercomputer can generate numerical computation result of over tera bytes. Even if using data compression technique, network bandwidth is not enough for data transmission through the Internet. In the next generation supercomputer, the data transmission problem will be obvious and we will not be able to handle post processing, which is an examination task using visualization software system after simulation, on outside computers. There are two ways to solve the problem. One is the data compression which is reduced data set composed by coarse volume data or coarse facet data using parallel computer environment [5–8]. The data set becomes much smaller than the original data and we can handle the post processing with interactivity. The other is the direct visualization on the supercomputer which is sequential post processing after numerical simulation [9–11]. However computer visualization as a post processing in the finite element analysis provides important user experience. From a point of this view, visualization on the computer could not provide experience well in the past. In addition, the architecture of the next generation supercomputer is just determined. In the background, the authors have developed scientific visualization library: LexADV_VSCG [9, 12–14] with high portability for any computer environment. In this study, several results are shown and a design and an actual implementation for the library through the visualized results are discussed.

2 LexADV_VSCG Library

2.1 Fundamental Ideas of Design and Implementation

Fundamental ideas of LexADV VSCG are as follows [3].

- Simple implementation for any environment: independent from OS.
- Multiple images to be one image using z-buffer: parallel processing.
- No other special hardware and libraries: independent from specific environment issue.
- Handling very fine resolution by well-designed software: key idea for ultra large scale data.
- Impressive image by well-designed algorithm and implementation: key idea using very fine resolution.

These ideas are important keys for high reliability on the future supercomputer. As mentioned above, we do not know the actual next generation supercomputer very well. In order to ensure those ideas, we have decided rendering processes conducted by software without any other software library. LexADV_VSCG is a simple library software which provides application programming interface for scientific visualization. In short, LexADV_VSCG is usually integrated into a simulation code and does not use any specific hardware accelerators and software libraries which are developed by other research group and companies.

Image data structure has additional several bits to represent depth information and dot product by surface normal and view direction. The structure possess extended z-buffer information. The advantages of the extended z-buffer are suited for parallel computers on which the images are rendered in parallel. The images are usually stored at the local node storage or stay in local memory. For example, if a supercomputer has totally 50,000 cores, FE mesh also is subdivided into 50,000 parts. View camera is fixed at 200 locations, 10,000,000 images are rendered at least. We should merge all of these images into 200 images in this case. As a result, the merging process needs huge data transmission between nodes. Merging tasks are sequentially processed as shown in Fig. 1. If the tasks are processed for 16 times, 2^{16} images can be merged into one.

Visualization technique of LOD (Level Of Detail) is a practical idea to deal with a very fine FE mesh. Simplicity of handling the visualized results is important. LOD requires several level of data set for one analysis model. In case of massively parallel data, we have to manage hundreds of thousands of data sets. Furthermore, very large scale FE data can be generated on only a supercomputer. If computation from modeling to visualization is conducted on a supercomputer, computing cost would be larger than we expected at first. In order to overcome these difficulties, we have decided to generate $10^5 \times 10^5$ pixels image in parallel and we just merge hundreds of thousands of images into one on the supercomputer.

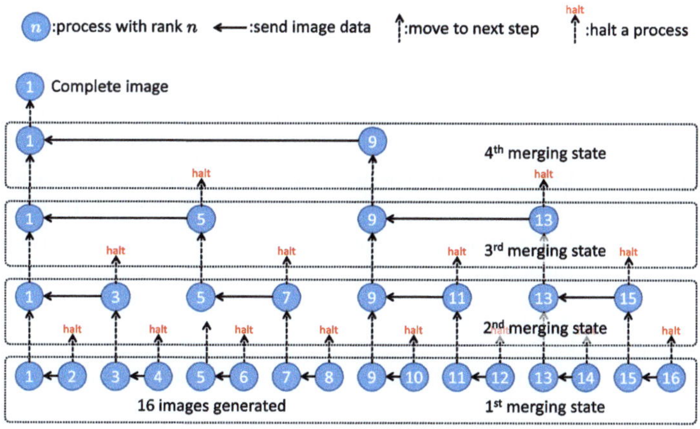

Fig. 1 Image merging process in parallel environment

2.2 Implementation and Functions

LexADV_VSCG is implemented in C language which is compatible with C99 standard. Any kind of compiler on supercomputers usually does not provide the latest

Table 1 Implemented fundamental functions for scientific visualization

	Functions	Explanation
vscg_draw_traingle functions	draw_triangle_line	Draw triangle line with same color and have no depth info
	draw_triangle_solid	Draw solid triangle with same color and have no depth info
	draw_triangle_line_with_depth (z-buffer)	Draw triangle line with color and have depth info
	draw_triangle_solid_with_depth (z-buffer)	Draw solid triangle with same color and have depth info
	draw_triangle_gradated_with_depth	Draw solid triangle with gradated color and have depth info
	draw_triangle_gradated_transparent	Draw transparent triangle with gradated color and have no depth info
vscg_XXXX utility functions	allocate_image	Allocate image buffer for drawing
	set_color_XXXX (setting default color)	Generate colors for basic colors and color legend
	vector_XXXX (vector operations)	2 and 3 dimensional vector operations
	accumrate_image (image operations)	Merge two or more z-buffer images into one image
vscg_draw_particle functions	draw_particle_solid (regular icosahedron)	Draw regular icosahedron as a sphere with depth info
	draw_particle_transparent_solid (regular icosahedron)	Draw transparent regular icosahedron as a sphere with depth info
	–	–
vscg_XXXX utility functions	draw_vector_with_arrow	Draw vector using arrow shape with depth info
	paint_color_legend	Paint color legend for physical quantity distribution
	paint_color_legend_2_physical_quantities	Paint color legend for 2 physical quantity distributions

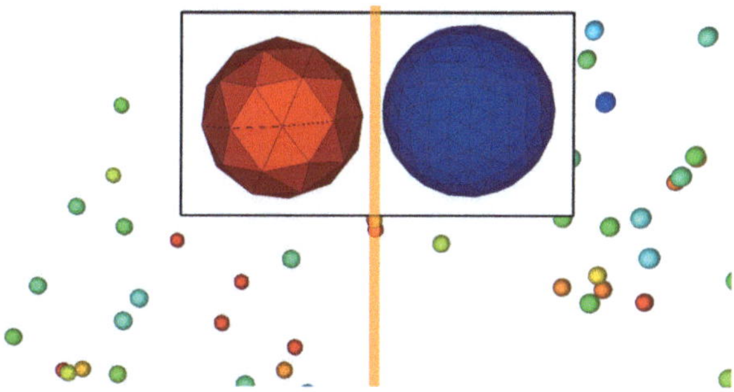

Fig. 2 Particle representation by a polyhedron: coarse polyhedron in *left side* and fine polyhedron in *right side*

standard, because reliability and performance are required to compilers. Since a reliability is strongly demanded in scientific visualization, LexADV_VSCG needs no other library except for standard C library defined in C99 standard. Functions in the library are shown in Table 1. The library provides very fundamental functions to draw and render triangles with gradated colors in accordance with specified physical quantity, i.e. von Mieses stress $\overline{\sigma}$ and principal stresses σ_1, σ_2 and σ_3 and so on. One of the target application of the library is particle based method, i.e. Moving Particle Semi-implicit (MPS) and Smoothed-Particle Hydrodynamics (SPH). A sphere is represented by a polyhedron which consists of 80 or 320 triangles. An example of particle visualization is shown in Fig. 2. Total numbers of polygons affect total time to get an image, however it is important for users to choose fine or coarse visualization result. In the very fine image, differences between fine and coarse polyhedron do not make large difference of both of images.

3 How to Get Interactivity Using the Library

In the development of LexADV_VSCG library, the objective is clearly limited to applications to very large scale fine mesh and huge number of particles. The library does not provide such usual techniques for high quality rendering image, which are ray casting and glow shading. As a result of these limitation, the implementation goes back to fundamental techniques by drawing triangle facet and shading on the flat triangle. However, huge number of fine facet and particle which represented by triangle facets, the quality of the image rendered by the library is enough for scientific visualization with the supercomputer.

Interactivity of visualization is one of the most important issue in the scientific post processing. Regretfully the latest supercomputer does not provide any interactive connection with user's computer. All of programs are controlled by job controlling system on the supercomputer. LexADV_VSCG library is able to generate very fine image with $10^5 \times 10^5$ pixels resolution. The image has enough information with regards to fine FE mesh. Usual image viewer can easily handle the large image by shrinking to appropriate size as an engineer needs. We require interactivity of

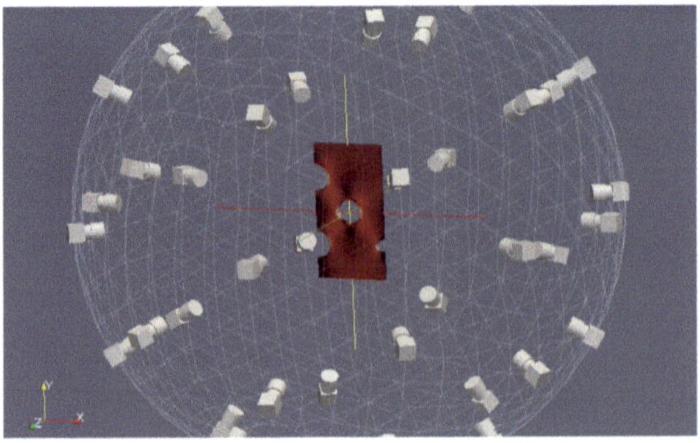

Fig. 3 Multiple view points for off-line interactive visualization

`{name}_{seq. num.}_{θ in deg.}_{φ in deg.}_{time step}.png`

- Name: arbitrary numbers and alphabets within 256 letters
- Sequential number: distinguishing analysis case by 8 digits
- θ and φ in degree: angle in degree , i.e. 01050 = 10.50°
- Time step: time for dynamic analysis by 6 digits

Example:
`MechanicalPart_00000001_01050_00050_000000.png`

Fig. 4 Notation rule for camera position in spherical surface

rotation of an object too. The system provides multiple viewpoints image generation as shown in Fig. 3. Simple spherical polar coordinate system and the description rule of the file name including the coordinate system are defined as described in Fig. 4. The rule is easily extended to other coordinate system, i.e. Euler's angle definition. There are two ways to describe rotation of an object in the library. First way directly describes rotation matrix. Second way is Rodrigues vector (k_1, k_2, k_3). Rodrigues' rotation formula is described as follows

$$\mathbf{K} = \begin{pmatrix} 0 & -k_3 & k_2 \\ k_3 & 0 & -k_1 \\ -k_2 & k_1 & 0 \end{pmatrix} \tag{1}$$

$$\mathbf{R} = \mathbf{I} + (\sin\theta)\mathbf{K} + (\mathbf{1} - \cos\theta)\mathbf{K}^2 \tag{2}$$

where θ is length of Rodrigues vector. Camera position is notated in spherical polar coordinates. The distance from the center of an object to camera is constant, which is 1. Amplitude θ is an angle between z-axis and radius vector. Amplitude φ is an angle between x-axis and projected vector of radius vector to x–y plane. In actual implementation, the camera is fixed at $(0, 0, -1)$. In order to describe the notation, camera positions should be calculated by

$$\begin{Bmatrix} x \\ y \\ z \end{Bmatrix} = \mathbf{R}^{-1} \begin{Bmatrix} 0 \\ 0 \\ -1 \end{Bmatrix} \tag{3}$$

$$\begin{cases} \cos\theta = \dfrac{z}{\sqrt{x^2+y^2+z^2}} & (0 \le \theta \le \pi) \\ \cos\varphi = \dfrac{x}{\sqrt{x^2+y^2}},\ \sin\varphi = \dfrac{y}{\sqrt{x^2+y^2}} & (0 \le \varphi \le 2\pi) \end{cases} \tag{4}$$

where x, y and z are camera position in orthogonal coordinate system; θ and φ are camera position in spherical polar coordinate system.

4 Results in Numerical Examples

4.1 Application to Finite Element Analysis in Structural Problem

Visualization techniques are well designed and developed using GPGPU hardware and those can present impressive artificial images. However, usual design process and analysis process does not demand for such an impressive image but accurate and easy to understand. From a practical point of view, the library supports polygon with flat shading including alpha blending which represent transparency effect.

Figure 5 shows an example of high resolution FE analysis image, which is resized image from $10^5 \times 10^5$ pixels. In the image, black lines with 1 pixel width are drawn for representing finite elements, however no element lines cannot be seen in Fig. 5a. The $10^5 \times 10^5$ pixels image shows element lines as shown in Fig. 5b. Figure 5b is magnified image from the $10^5 \times 10^5$ pixels image. Gradated color is very smoothly distributed in spite of linear interpolation of gradated color. Engineers want to see finite elements even if a very large scale mesh. Since manipulations of 2-dimensional image is processed very fast, magnification and reduction processes are done for a short time.

4.2 Application to Finite Element Analysis for Cross Section Generation

One of the important requirement is cross section generation with gradated color for a large scale problem. In the large scale finite element analysis, the complexity become higher and higher. For example, Yoshimura et al. [15, 16], analyzed full-scale nuclear power plant simulation for seismic dynamics response. In the analysis, the plant model was created with 200 million D.O.F. in detail. For such an application, we have to watch arbitrary sections very carefully for evaluation of FE analysis result. A

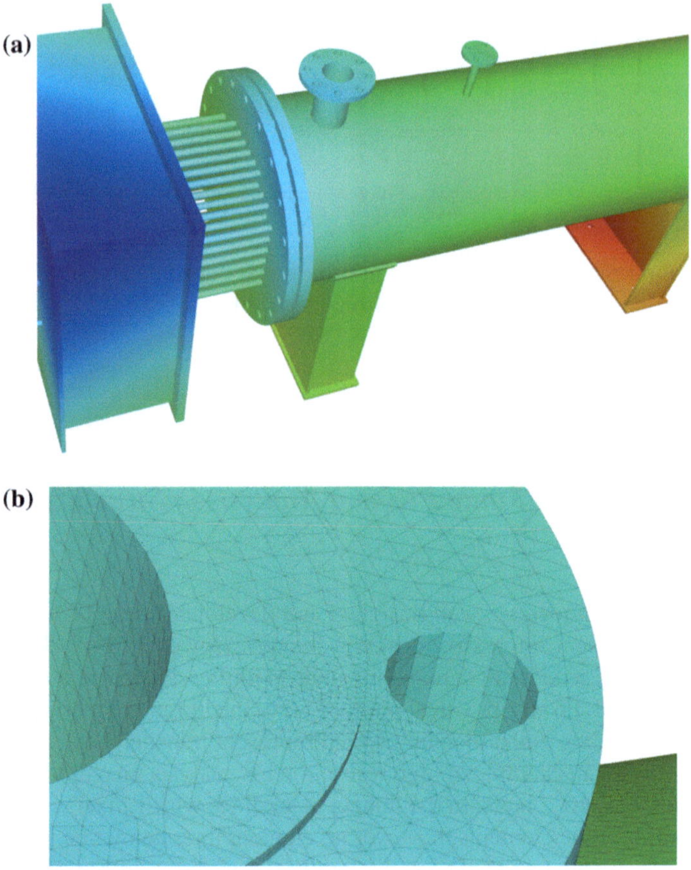

Fig. 5 Visualized results for finite element model with $10^5 \times 10^5$ pixel resolution: **a** whole view and **b** magnified view

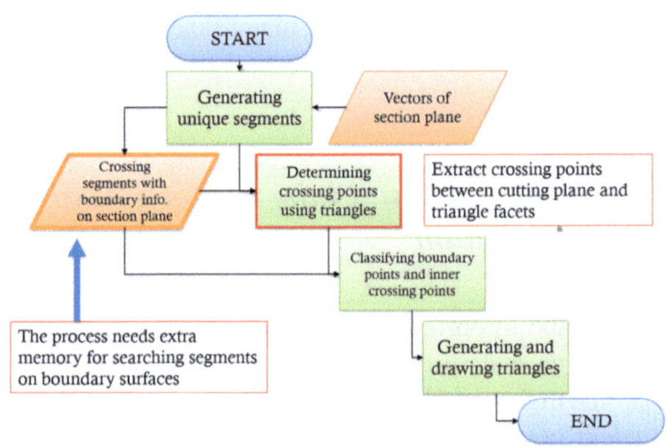

Fig. 6 Flow chart of cross section generation for tetrahedrons and triangles of finite element

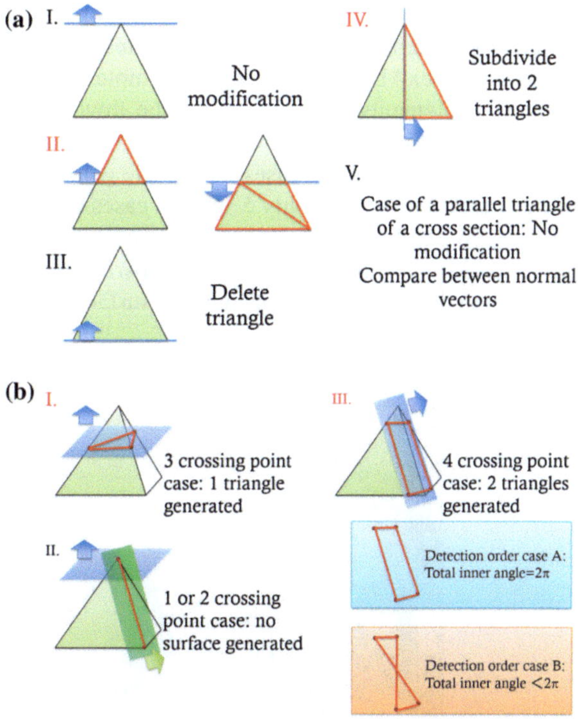

Fig. 7 Section generation on triangle surface and tetrahedral volume: **a** Triangle surface subdivision and **b** Tetrahedral volume subdivision

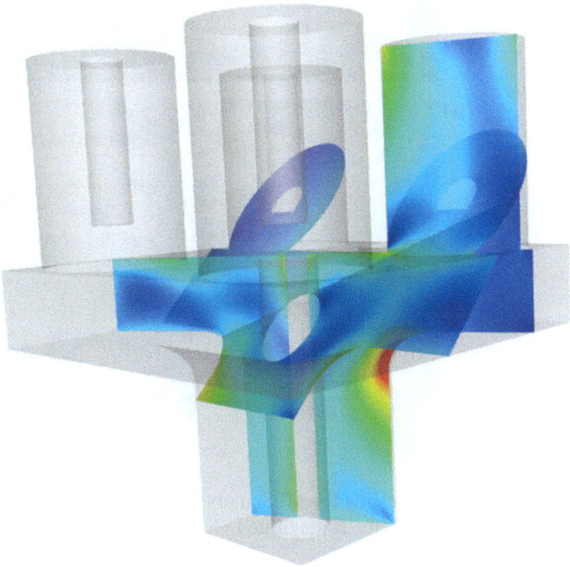

Fig. 8 Two sections and transparent model surface with equivalent stress distribution

tetrahedral element is utilized for large scale analysis, because fully automatic mesh generation system is only established in tetrahedral element mesh generation. For this reason, LexADV_VSCG employ simple algorithm of section generation from tetrahedral elements. Figure 6 shows the flow chart of section surface generation algorithm and Fig. 7 shows how to subdivide triangle surface and tetrahedral volume. There are only 15 unique cases of triangle generation in Marching cube algorithm. In the LexADV_VSCG, only 5 unique cases of triangle generation for tetrahedral volume are employed. Several cross sections can be simultaneously generated by the algorithm as shown in Fig. 8. Transparent model surface is rendered by alpha blending to draw cross sections and the surface at the same time.

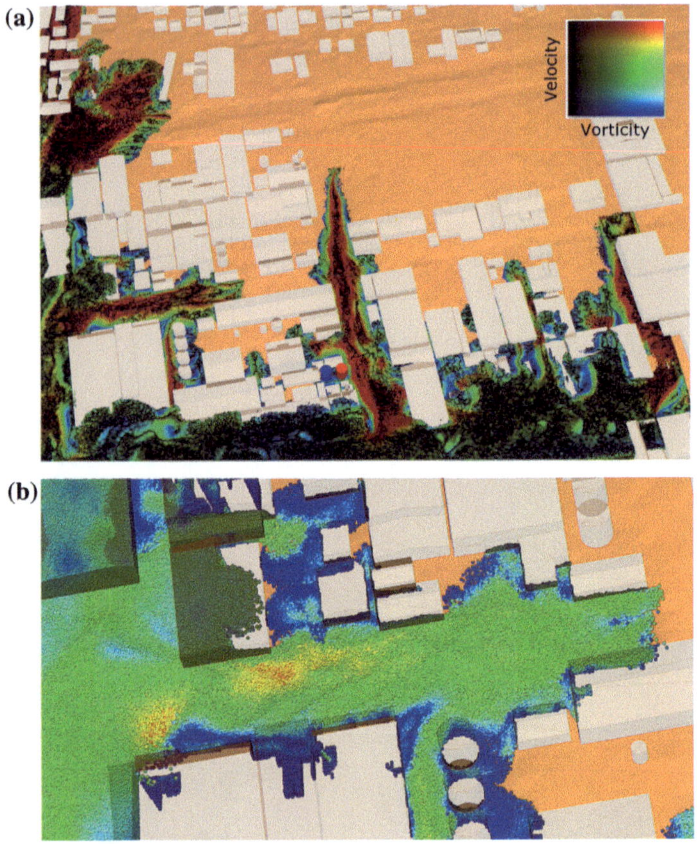

Fig. 9 Visualized results for MPS analysis with 30,000 × 20,000 pixels resolution: **a** whole view of velocity and vorticity, **b** magnified view of velocity field at the same time step

4.3 Application to Particle Based Method in Fluid Dynamics

Figure 9 shows another example of very high resolution analysis by explicit MPS method. The visualization is conducted using transparent object and solid colored particles. Figure 9a shows 16,000 × 12,800 pixels image which is enough resolution to examine fluid flow velocity and vorticity. Figure 9b is magnified image of velocity from the original image. A large number of particles forms the 3-dimensional configuration of fluid surface in Fig. 9b, since the sufficient resolution of images keeps shadow of each particle and automatically makes the surface formed.

5 Conclusions

We have developed library software to visualize scientific computation results on the next generation supercomputer. LexADV_VSCG generates very fine images on the parallel environment efficiently. Interactivity is one of the important issue for examination of analysis result. We have also proposed interactive operation way by simple representation of spherical polar coordinate system. In particular structural analysis engineers usually examine the result using interactive visualization software. Basic visualization techniques are already implemented in the library. Huge amount of data, cannot be downloaded through network, is computed by supercomputers. Automated process to obtain the best visualization results is required for the next generation visualization software. The library just provides fundamental functions and reliability. The library is now developed in progress and is evaluated for more modifications to ensure reliability.

Acknowledgments This research was supported in the Basic Research Programs: Development of System Software Technologies for post-Peta Scale High Performance Computing by CREST, JST.

References

1. H. Nakamura, I. Fujishiro, Y. Takeshima, Towards optimizing local feature metric for simplifying colored interval volumes, in *Proceedings of Work in Progress, IEEE Visualization 2000* (Salt Lake City, 2000), pp. 59–66
2. I. Fujishiro et al., Parallel visualization of gigabyte datasets in GeoFEM, *Concurrency and Computation: Practice and Experience*, vol. 14 (Wiley, Chichester, 2002), pp. 521–530
3. H. Okuda et al., Parallel finite element analysis platform for the Earth simulator: GeoFEM, in *Proceedings of 3rd International Working Group Meeting* (2003), p. 6
4. J. Dongarra et al., The international exascale software project roadmap. Int. J. High Perform. Comput. Appl. **25**, 3–60 (2011)
5. K. Ma et al., In-situ processing and visualization for ultrascale simulations. J. Phys.: Conf. Ser. **78**(012043), 10 (2007)
6. K. Ma, In situ visualization at extreme scale: challenges and opportunities. Comput. Graph. Appl., IEEE **29**(6), 14–19 (2009)

7. N. Fabian, K. Moreland, The paraview coprocessing library: a scalable, general purpose in situ visualization library. *IEEE Symposium on Large Data Analysis and Visualization, LDAV*, IEEE, 2011, pp. 89–96
8. A. Artigues et al., Scientific big data visualization: a coupled tools approach. Int. J. Supercomput. Front. Innov. **1**(3), 4–18 (2014)
9. Y. Wada et al., Development of high resolution visualization library for very large scale analysis, in *Proceedings of 18th JSCES Conference*, vol. 18, 2 p. (2013)
10. A. Kageyama, T. Yamada, An approach to exascale visualization: interactive viewing of in-situ visualization. Comput. Phys. Commun. **185**(1), 79–85 (2014)
11. J. Ahrens et al., An image-based approach to extreme scale in situ visualization and analysis, in *SC '14 Proceedings of the International Conference for High Performance Computing, Networking, Storage and Analysis*, pp. 424–434 (2014)
12. Y. Wada et al., Development of high resolution visualization library for very large scale analysis, in *Proceedings of JSME-KSME Joint Symposium on CM & CAE 2014*, 4 p. (2014)
13. Y. Wada et al., Advanced high resolution visualization library VSCG for very large scale analysis, in *Proceedings of 19th JSCES Conference*, vol. 19, 2 p. (2014)
14. Y. Wada et al., Implementation of polygon-based section generation for high resolution visualization library LexADV_VSCG, in *Proceedings of 20th JSCES Conference*, vol. 20, 2 p. (2015)
15. S. Yoshimura et al., Full scale seismic response simulation of nuclear power plant using K-computer, in *Abstracts of COMPDYN2013, 4th International Conference on Computational Methods in Structural Dynamics and Earthquake Engineering* (2013), http://www.eccomasproceedings.org/cs2013/pdf/1734.pdf
16. S. Yoshimura et al., Petascale simulation based investigation on structural integrity of nuclear power plant attacked by strong earthquake, in *Abstracts of 11th World Congress on Computational Mechanics (WCCM XI)* (2014), http://www.wccm-eccm-ecfd2014.org/frontal/PLENARY/Abstracts/SYoshimura.pdf

Drawing Curves

Toshio Oshima

Abstract We propose a method to determine piecewise cubic Bézier curves passing through given points. Our main purpose is to draw accurate graphs of mathematical functions with smaller data. A program drawing such graphs using our method is realized in a computer algebra and outputs the graphs in a source file of TEX and then transforms it into a PDF file. Our method is also useful for numerical calculation of a given area enclosed by a curve and for numerical integration of functions.

Keywords Bézier curve · Cubic spline · Computer algebra · Risa/Asir · TEX · TikZ · 3D graph · Numerical integration

1 Introduction

Since the last year I have a class of calculus in my university and show graphs of functions such as $f(x, y) = x^2 - y^2$. I have been developing a library os_muldif.rr [1] of a computer algebra Risa/Asir [2] to realize my research explained in [3] and then I added some functions in the library for such educational purpose including calculus, linear algebra and elementary number theory. The library is an open source and can be equally executed by a personal computer with any one of the operating systems Windows, Mac and UNIX.

In fact, a function in the library executes the procedure in Fig. 1 to get the graphs. Since the PDF file supports cubic Bézier curves, the size of the PDF file obtained in the procedure is usually small and it is independent of the final resolution of the graph.

T. Oshima (✉)
Josai University, 2-3-20 Hirakawacho, Chiyodaku, Tokyo 102-0093, Japan
e-mail: t-oshima@josai.ac.jp

Y. Dobashi and H. Ochiai (eds.), *Mathematical Progress in Expressive Image Synthesis III*, Mathematics for Industry 24,
DOI 10.1007/978-981-10-1076-7_11

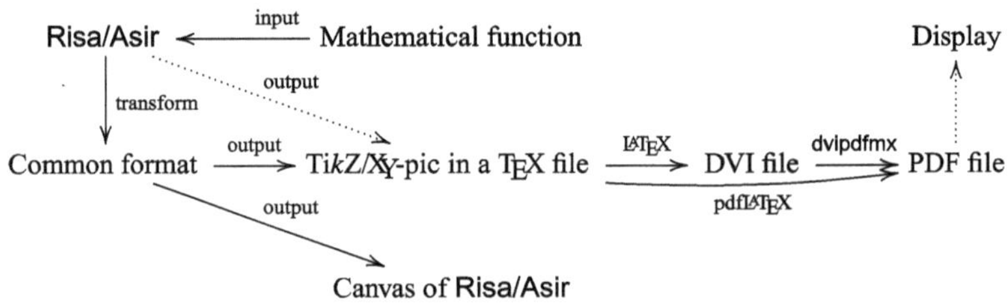

Fig. 1 Procedure

2 Curves

Consider a curve

$$C : [a, b] \ni t \mapsto \gamma(t) = \big(x(t), y(t)\big) \in \mathbb{R}^2. \tag{1}$$

We choose points in $[a, b]$, namely, $P_j = \gamma(t_j) \in C$ with $a = t_0 < t_1 < t_2 < \cdots < t_N = b$ and draw a certain curve C' starting from P_0, exactly passing through $P_1, \ldots, P_{N-1}$ in this order and ending at P_N. We request the following conditions.

- C' is determined only by $\{P_0, P_1, \ldots, P_N\}$.
- C' is a good approximation of C and it is free from its final resolution in drawing.
- Smaller size of data (i.e. the number N) and an output in a popular format are desirable.
- The curve can be described in a usual TEX sourceTEX file.

One of the way to realize it is to connect the points by cubic Bézier curves and use Ti*k*Z and/or XY-pic which are in a package of a TEX system (cf. Fig. 1).

2.1 Smooth Curves

A Bézier curve of degree n is

$$[0, 1] \ni t \mapsto P(t) = P(B_0, \ldots, B_n; t) = \sum_{i=0}^{n} \binom{n}{i} t^i (1-t)^{n-i} B_i \tag{2}$$

determined by $(n + 1)$ points $B_0, \ldots, B_n$.

Note that $P(B, B'; t)$ is the point internally dividing the line segment BB' by $t : 1 - t$. Since $P(B_0, \ldots, B_n; t) = P(P(B_0, B_1; t), P(B_1, B_2; t), \ldots, P(B_{n-1}, B_n; t); t)$, the point $P(t)$ is geometrically described. For example, the **cubic Bézier curve** is

$$\begin{aligned} P(t) &= P(B_0, B_1, B_2, B_3; t) = P\big(P(B_0, B_1; t), P(B_1, B_2; t), P(B_2, B_3; t); t\big) \\ &= P\Big(P\big(P(B_0, B_1; t), P(B_1, B_2; t); t\big), P\big(P(B_1, B_2; t), P(B_2, B_3; t); t\big); t\Big). \end{aligned}$$

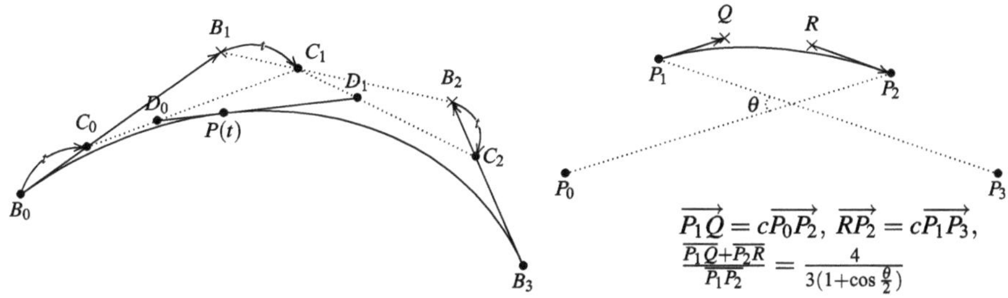

Fig. 2 Cubic Bézier curves

The curve starts from B_0 to the direction $\overrightarrow{B_0B_1}$ and ends at B_3 to the direction $\overrightarrow{B_2B_3}$. It does not necessarily pass through B_1 nor B_2.

Consider a curve C passing through P_0, P_1, P_2, P_3 in this order. We simulate the curve segment of C connecting P_1 to P_2 by the cubic Bézier curve $P(P_1, Q, R, P_2; t)$ with the control points Q and R defined in Fig. 2. The number c is determined by

$$c = \frac{4\overline{P_1P_2}}{3(\overline{P_0P_2} + \overline{P_1P_3})} \frac{1}{1 + \sqrt{\frac{1}{2}\left(1 + \frac{(\overrightarrow{P_0P_2}, \overrightarrow{P_1P_3})}{\overline{P_0P_2} \cdot \overline{P_1P_3}}\right)}}. \tag{3}$$

To explain (3) we assume that $\overline{P_0P_1} = \overline{P_1P_2} = \overline{P_2P_3}$ and moreover that $P_0, \ldots, P_3$ are on a circle with the center O. We define a Bézier curve with the control points Q and R which approximates the arc connecting P_1 and P_2. Putting $\angle P_1OP_2 = \theta$, $\overline{OP_1} = r$, $\overline{P_1Q} = \overline{P_2R} = a$, the point T on the Bézier curve corresponding to $t = \frac{1}{2}$ is given as follows under a suitable coordinate system.

$$\begin{aligned}
&O : (0, 0),\ P_1 : \left(r\cos\tfrac{\theta}{2}, r\sin\tfrac{\theta}{2}\right),\ P_2 : \left(r\cos\tfrac{\theta}{2}, -r\sin\tfrac{\theta}{2}\right)\\
&Q : \left(r\cos\tfrac{\theta}{2} + a\cos\tfrac{\theta-\pi}{2}, r\sin\tfrac{\theta}{2} + a\sin\tfrac{\theta-\pi}{2}\right)\\
&\quad = \left(r\cos\tfrac{\theta}{2} + a\sin\tfrac{\theta}{2}, r\sin\tfrac{\theta}{2} - a\cos\tfrac{\theta}{2}\right)\\
&R : \left(r\cos\tfrac{\theta}{2} + a\sin\tfrac{\theta}{2}, -r\sin\tfrac{\theta}{2} + a\cos\tfrac{\theta}{2}\right)\\
&T : \left(r\cos\tfrac{\theta}{2} + \tfrac{3}{4}a\sin\tfrac{\theta}{2}, 0\right)
\end{aligned}$$

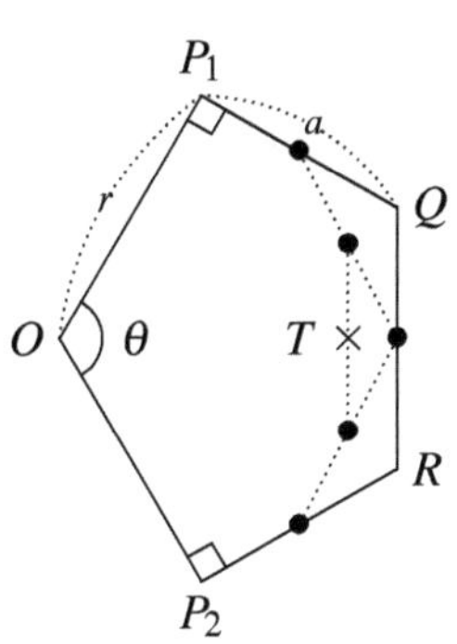

Put $\overline{OT} = \overline{OP_1}$ to approximate the arc. Then

$$r\cos\tfrac{\theta}{2} + \tfrac{3}{4}a\sin\tfrac{\theta}{2} = r$$

and therefore

$$a = \frac{4}{3}\frac{1-\cos\frac{\theta}{2}}{\sin\frac{\theta}{2}}r = \frac{4}{3}\frac{\sin\frac{\theta}{2}}{1+\cos\frac{\theta}{2}}r.$$

In this case we have

$$\begin{aligned}
Q &: \left(r\cos\tfrac{\theta}{2} + \frac{4}{3}\frac{1-\cos\frac{\theta}{2}}{\sin\frac{\theta}{2}}\sin\tfrac{\theta}{2}r,\ r\sin\tfrac{\theta}{2} - \frac{4}{3}\frac{\sin\frac{\theta}{2}}{1+\cos\frac{\theta}{2}}\cos\tfrac{\theta}{2}r\right)\\
&= \left(\left(\tfrac{4}{3} - \tfrac{1}{3}\cos\tfrac{\theta}{2}\right)r,\ \left(1 - \tfrac{1}{3}\cos\tfrac{\theta}{2}\right)\frac{\sin\frac{\theta}{2}}{1+\cos\frac{\theta}{2}}r\right),\\
\frac{\overline{P_1Q}}{\overline{P_1P_2}} &= \frac{4}{3}\frac{\sin\frac{\theta}{2}}{1+\cos\frac{\theta}{2}}\frac{1}{2\sin\frac{\theta}{2}} = \frac{2}{3(1+\cos\frac{\theta}{2})}.
\end{aligned} \tag{4}$$

Put $r = 1$ and $c = \cos\frac{\theta}{2}$. We examine the distance between O and the point

$$B(t) = \big(x(t), y(t)\big) = P_1(1-t)^3 + 3Qt(1-t)^2 + 3Rt^2(1-t) + P_2t^3$$

on the Bézier curve. Denoting $t = s + \frac{1}{2}$, we have

$$\begin{aligned}
L(s) &:= x\left(s+\tfrac{1}{2}\right)^2 + y\left(s+\tfrac{1}{2}\right)^2\\
&= \frac{16(1-c)^3}{1+c}s^6 - \frac{8(1-c)^3}{1+c}s^4 + \frac{(1-c)^3}{1+c}s^2 + 1\\
&= \tfrac{(1-c)^3}{1+c}s^2(4s^2-1)^2 + 1
\end{aligned}$$

and when $0 \le s \le \frac{1}{2}$,

$$\sqrt[3]{8s^2(1-4s^2)^2} \ge \frac{8s^2 + (1-4s^2) + (1-4s^2)}{3} = \frac{2}{3}.$$

The equality in the above holds if and only if $8s^2 = 1 - 4s^2$, namely, $s^2 = \frac{1}{12}$. Hence $L(s)$ with $|s| \le \frac{1}{2}$ takes the minimal value 1 when $s = 0,\ \pm\frac{1}{2}$ and the maximal value when $s = \pm\frac{1}{2\sqrt{3}}$.

$$L\left(\pm\tfrac{1}{2\sqrt{3}}\right) - 1 = \frac{1}{27}\frac{(1-c)^3}{1+c},$$

$$\sqrt{L\left(\pm\tfrac{1}{2\sqrt{3}}\right)} - 1 \fallingdotseq \frac{1}{54}\frac{\left(1-\cos\frac{\theta}{2}\right)^3}{1+\cos\frac{\theta}{2}} \fallingdotseq \begin{cases} \frac{1}{648} & \left(\theta = \frac{2\pi}{3} = 120°\right), \\ \frac{1}{3668} & \left(\theta = \frac{\pi}{2} = 90°\right), \\ \frac{1}{41900} & \left(\theta = \frac{\pi}{3} = 60°\right), \\ \frac{1}{235541} & \left(\theta = \frac{\pi}{4} = 45°\right), \\ \frac{1}{2683400} & \left(\theta = \frac{\pi}{6} = 30°\right). \end{cases} \tag{5}$$

In view of (4), we determine that the segment between P_1 and P_2 in the curve interpolating general P_0, P_1, P_2, P_3 is the cubic Bézier curve with the control points Q and R so that

$$\overrightarrow{P_1Q} = c\overrightarrow{P_0P_2}, \quad \overrightarrow{P_2R} = c\overrightarrow{P_3P_1}, \tag{6}$$

$$\frac{\overline{P_1Q} + \overline{P_2R}}{\overline{P_1P_2}} = \frac{4}{3\left(1+\cos\frac{\theta}{2}\right)}. \tag{7}$$

Thus

$$\cos\theta = \frac{(\overrightarrow{P_0P_2}, \overrightarrow{P_1P_3})}{\overline{P_0P_2}\cdot\overline{P_1P_3}}, \quad \cos\frac{\theta}{2} = \sqrt{\frac{1+\cos\theta}{2}} = \sqrt{\frac{1}{2}\left(1 + \frac{(\overrightarrow{P_0P_2}, \overrightarrow{P_1P_3})}{\overline{P_0P_2}\cdot\overline{P_1P_3}}\right)}$$

$$\text{and}\quad \frac{c\overline{P_0P_2} + c\overline{P_1P_3}}{\overline{P_1P_2}} = \frac{4}{3\left(1+\cos\frac{\theta}{2}\right)}$$

and therefore we have (3).

The cubic Bézier curve is given by

$$\begin{aligned} B(t) &= P_1(1-t)^3 + 3Qt(1-t)^2 + 3Rt^2(1-t) + P_2t^3 \\ &= (-P_1 + 3Q - 3R + P_2)t^3 + (3P_1 - 6Q + 3R)t^2 + (-3P_1 + 3Q)t + P_1. \end{aligned}$$

The Catmull-Rom spline curve is defined by

$$\begin{aligned} C(t) = &\left(-\tfrac{1}{2}P_0 + \tfrac{3}{2}P_1 - \tfrac{3}{2}P_2 + \tfrac{1}{2}P_3\right)t^3 + \left(P_0 - \tfrac{5}{2}P_1 + 2P_2 + \tfrac{1}{2}P_3\right)t^2 \\ &+ \left(-\tfrac{1}{2}P_0 + \tfrac{1}{2}P_2\right)t + P_1 \end{aligned}$$

and therefore the corresponding control points Q and R in this case are defined by

$$\begin{cases} Q = P_1 + \frac{1}{6}(P_2 - P_0), \\ R = P_2 + \frac{1}{6}(P_3 - P_1), \end{cases}$$

which means that we fix $c = \frac{1}{6}$ in (6).

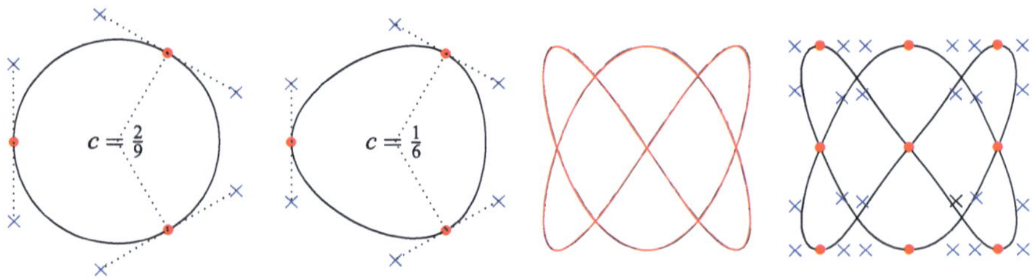

$N = 3$ (3 points) Catmull-Rom spline $N = 24, N = 96$ $N = 12$ (9 points)

Fig. 3 Bézier curves

In our case, the relative error $\left|\frac{\overline{OB(t)}}{\overline{OP_1}} - 1\right|$ is less than $\frac{1}{640}$ (resp. $\frac{1}{3600}$) if $\angle P_1OP_2 \leq 120°$ (resp. $\leq 90°$). Note that a Bézier curve never coincides with an exact arc.

For a closed curve C passing through points $R_0, R_1, \ldots, R_N = R_0$ in this order we draw a curve segment between R_j and R_{j+1} by putting $P_i = R_{i+j-1}$ for $i = 0, 1, 2$ and 3 as in the above and $R_{\nu \pm N} = R_\nu$ $(\nu = 1, \ldots, N)$. Then the resulting curve C' we draw is a smooth closed curve (of class C^1) which simulates C.

When the number c is fixed to be $\frac{1}{6}$ in (6), the corresponding curve is known as the (uniform) Catmull-Rom spline curve (cf. [4]). It is invariant under affine transformations and our curve is invariant under conformal affine transformations.

The following first example in Fig. 3 is the curve drawn by the three points $(\cos t, \sin t)$ with $t = \pm\frac{\pi}{3}, \pi$ indicated by •. The other 6 points calculated by using (3) are indicated by ×. In the final PDF file the positions of these 9 points are only written and the real rendering of the Bézier curve is done by a viewer of the file and therefore the size of the PDF file is small. The second example is the (uniform/centripetal) Catmull-Rom spline curve passing through these three points.

The other examples in Fig. 3 are the Lissajous curve $\gamma(t) = (\sin 2t, \sin 3t)$ drawn by the points corresponding to $t = \frac{2\pi j}{N}$ for $j = 0, \ldots, N$.

If the points $P_j = \gamma(t_j)$ are not suitably chosen, the resulting curve drawn by the points may be not good. Even in this case our curve is better than the corresponding Catmull-Rom spline curve as in the following example.

Suppose we draw a graph of the parabola defined by $y = x^2$. Taking the points on the curve $\gamma(t) = (t, t^2)$ corresponding to $t = -2, -1, 0, 0.2, 1, 2$, we draw curve for $-1 \leq t \leq 1$ by these 6 points in Fig. 4.

To avoid a singularity or a loop in a Bézier segment, a generalization of Catmul-Rom spline is introduced (cf. [5]):

$$\gamma(t) = \frac{t_2 - t}{t_2 - t_1} B_1 + \frac{t - t_1}{t_2 - t_1} B_2 \qquad (t \in [t_1, t_2]),$$

$$B_1 = \frac{t_2 - t}{t_2 - t_0} B_1 + \frac{t - t_0}{t_2 - t_0} B_2, \quad B_1 = \frac{t_3 - t}{t_3 - t_1} B_1 + \frac{t - t_1}{t_3 - t_1} B_2,$$

$$A_1 = \frac{t_1 - t}{t_1 - t_0} P_0 + \frac{t - t_0}{t_1 - t_0} P_1, \quad A_2 = \frac{t_2 - t}{t_2 - t_1} P_1 + \frac{t - t_1}{t_2 - t_1} P_2,$$

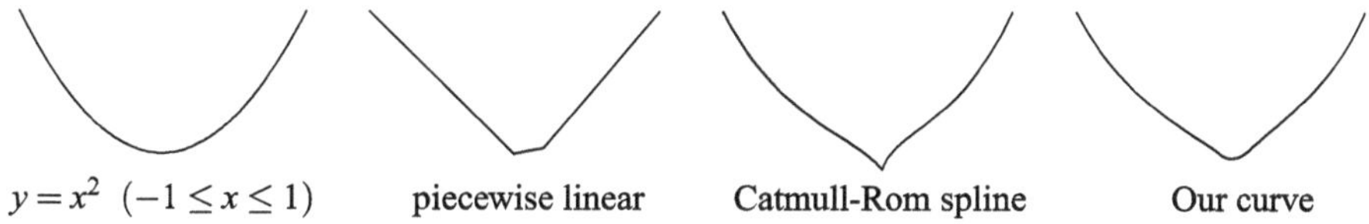

Fig. 4 Parabola

$$A_3 = \frac{t_3 - t}{t_3 - t_2} P_2 + \frac{t - t_2}{t_3 - t_2} P_3, \quad t_j = (\overline{P_{j-1}P_j})^{\alpha} + t_{j-1} \qquad (j = 1,\ 2,\ 3).$$

If $\alpha = 0$, the above curve equals the standard (uniform) Catmul-Rom spline. When $\alpha = 1$, the curve is called chordal Catmul-Rom spline. When $\alpha = 0.5$, the curve is called centripetal Catmull-Rom spline and has more desirable properties compared to the original one (cf. [6]). It will not form loop nor cusp within a curve segment.

But these Catmul-Rom splines produce the same result as in Fig. 3 for the points equally distributed on a circle because $\overline{P_{j-1}P_j}$ does not depend on j.

2.2 *Singularities*

We consider a curve $\gamma(t)$ $(t \in [a, b])$ which has singular points or discontinuous points. We assume that the curve is a finite union of smooth curves but we do not know the singular points of the curve.

First we choose points $P_j = \gamma(t_j)$ with $t_0 = a < t_1 < \cdots < t_N = b$ on the curve. We put $t_j = a + \frac{j(b-a)}{N}$ in most cases (or as default)[1].

For every j, add the point $\gamma(\frac{t_{j-1}+t_j}{2})$ if

- $\frac{(\overrightarrow{P_{j-2}P_{j-1}}, \overrightarrow{P_{j-1}P_j})}{\overline{P_{j-2}P_{j-1}} \cdot \overline{P_{j-1}P_j}} < C_1$ or $\frac{(\overrightarrow{P_{j-1}P_j}, \overrightarrow{P_jP_{j+1}})}{\overline{P_{j-1}P_j} \cdot \overline{P_jP_{j+1}}} < C_1$
  or
- $\overline{P_{j-1}P_j} > C_2$

Repeat the above up to m times,

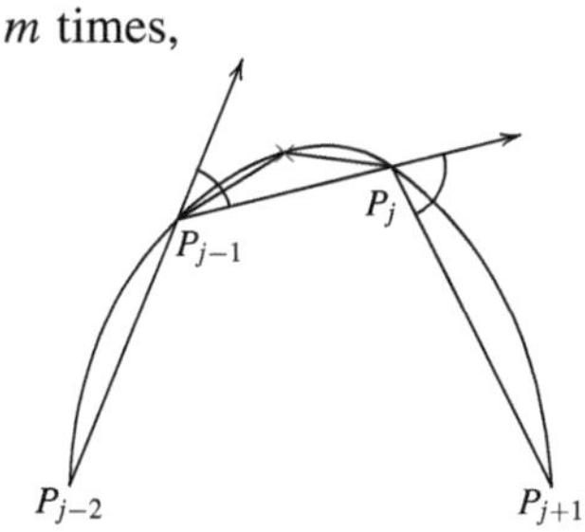

[1] Moreover if the curve is defined outside $[a, b]$, we use the points P_{-1} and P_{N+1} to define Bézier curves).

If the length $\overline{P_{j-1}P_j}$ still exceeds a given threshold value C_2 after this procedure, we cut our curve between two points P_{j-1} and P_j.

The default threshold values are $C_1 = \cos 30°$, $C_2 = \frac{\text{diameter of Window}}{16}$ and $m = 4$.

We examine the graph of the function

$$y = |2\sin x| - \big[|2\sin x|\big] \quad (0 \le x \le 5).$$

Here for a real number t, $\big[t\big]$ denotes the largest integer which does not exceed t.

Note that this function is discontinuous at $x = \frac{\pi}{6}, \frac{5\pi}{6}, \frac{7\pi}{6}$ and not smooth at $x = \pi$.

If we do not care the singularities, we have Fig. 5.

The procedure explained in this subsection gives Fig. 6 and the number of segments of Bésier curves increases from 32 to 70.

The graph of the absolute value of Riemann's zeta function $\zeta(z)$ for $\operatorname{Re} z = \frac{1}{2}$ is given in Fig. 7. Risa/Asir takes less than a second to get it in a PDF file.

Fig. 5 $y = \big||2\sin x| - \big[|2\sin x|\big]\big|$ $(0 \le x \le 5, m = 0 \text{ and } N = 32)$

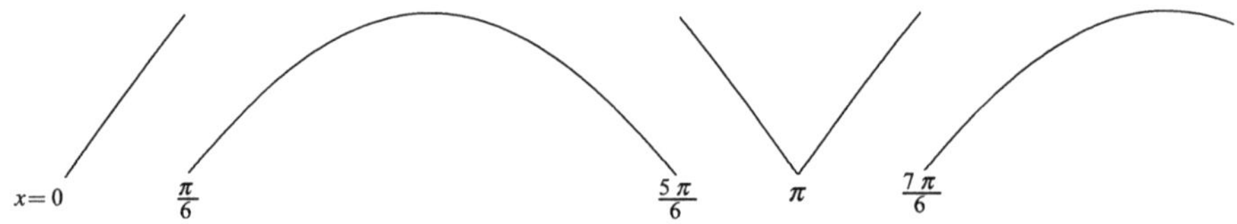

Fig. 6 $y = \big||2\sin x| - \big[|2\sin x|\big]\big|$ $(0 \le x \le 5, m = 4 \text{ and } N = 32 \to 70)$

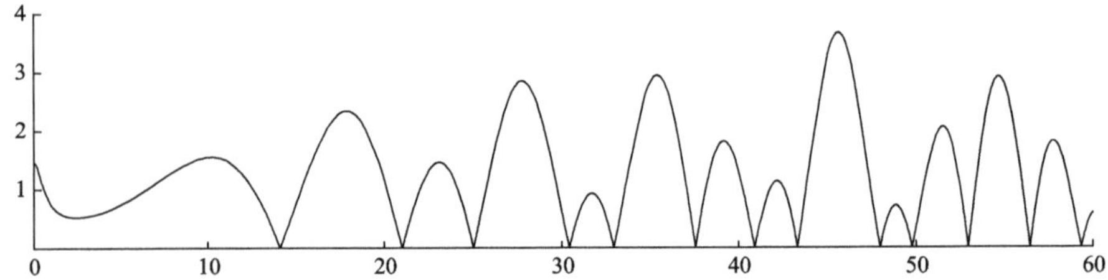

Fig. 7 $y = \left|\zeta\left(\frac{1}{2} + x\sqrt{-1}\right)\right|$ $(m = 6 \text{ and } N = 96 \to 355)$

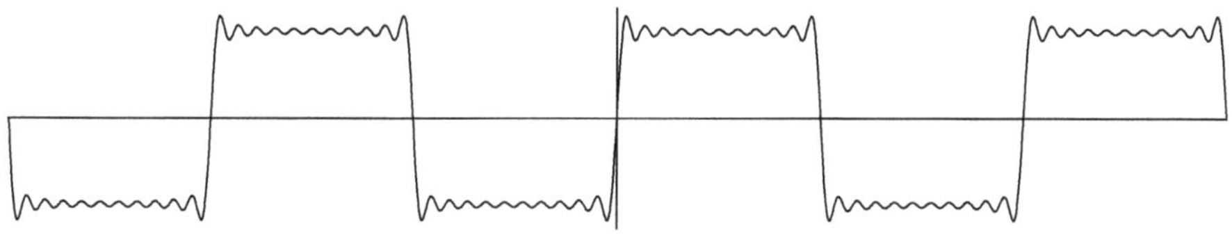

Fig. 8 Fourier series $(m = 6 \text{ and } N = 192 \to 1020)$

The final example in this subsection is the finite Fourier series

$$y = \sin x + \tfrac{1}{3}\sin\tfrac{x}{3} + \tfrac{1}{5}\sin\tfrac{x}{5} + \cdots + \tfrac{1}{21}\sin\tfrac{x}{21}$$

which approximates a square wave in Fig. 8.

3 Applications

3.1 *Circles, Arcs and Ovals*

The relative error of our approximation of an arc by a cubic Bézier curve becomes smaller when its central angle becomes smaller. If the angle is smaller than 120° (resp. 90°), then it is smaller than 0.16 % (resp. 0.028 %) as is shown in the previous section. The relative error here is measured by the distance from the center of the circle containing the arc.

Hence the central angle of an arc is not large, it is sufficient for us to approximates it by a single cubic Bézier curve or at most three cubic segments for most purposes.

Moreover since the Bézier curve is compatible with affine transformations, we can also draw an oval and an arc of an oval with the same accuracy by using an affine transformation of our approximation of a circle or an arc of a circle. These are realized in [1].

3.2 *Integration*

The area enclosed by a curve is numerically calculated by our approximation since an area enclosed by a curve with cubic Bézier segments is easily calculated (Fig. 8).

Suppose an area is enclosed by segments of cubic Bézier curves

$$[0, 1] \ni t \mapsto \gamma_j(t) = \big(x_j(t), y_j(t)\big) \qquad (j = 0, \ldots, N).$$

Then the absolute value of the line integral

$$I(\gamma) = \sum_{j=0}^{N} \int_0^1 y_j(t)\, dx_j(t) = \sum_{j=0}^{N} \int_0^1 x_j'(t) \cdot y_j(t)\, dt$$

gives the area. Here $x_j'(t) \cdot y_j(t)$ are polynomials of degree 5 and therefore the above value is easily calculated.

If the curve is an approximation of the graph of $y = f(x)$ with $x \in [a, b]$, the above value is an approximation of $\int_a^b f(x)\, dx$.

In the following table we show examples of the relative errors of the numerical integrations using this method. In the table, circle and cardioid are parametrized by

$$(\cos\theta, \sin\theta) \text{ and } \big((1+\cos\theta)\cos\theta, (1+\cos\theta)\sin\theta\big),$$

respectively. For example, in the case of cardioid in the table, "32 parts" means that the cardioid is approximated by 32 cubic Bézier segments determined only by the points $\big((1 + \cos\theta_j)\cos\theta_j, (1 + \cos\theta_j)\sin\theta_j\big)$ with $\theta_j = \frac{j\pi}{16} - \pi$ and $j = 0, 1, \dots, 32$ and the approximated area is calculated by the segments.

Integrations using Bézier curves

Curve	Interval	16 parts	32 parts	96 parts	384 parts	1536 parts
Circle	$0 \le \theta \le 2\pi$	6.8×10^{-8}	1.1×10^{-9}	1.5×10^{-12}	3.2×10^{-17}	8.7×10^{-20}
Cardioid	$-\pi \le \theta \le \pi$	5.4×10^{-4}	3.1×10^{-5}	3.8×10^{-7}	1.5×10^{-9}	5.8×10^{-12}
$x \sin x$	$0 \le x \le \pi$	2.9×10^{-4}	1.8×10^{-6}	2.2×10^{-8}	8.7×10^{-11}	3.4×10^{-13}
$\frac{\sin x}{x}$	$0 < x \le \pi$	1.5×10^{-6}	9.5×10^{-8}	1.2×10^{-9}	4.6×10^{-12}	1.7×10^{-14}
$\frac{1}{x^2+1}$	$-\infty < x < \infty$	1.3×10^{-5}	1.3×10^{-7}	8.5×10^{-10}	4.7×10^{-12}	2.1×10^{-14}
e^{-x^2}	$-\infty < x < \infty$	7.1×10^{-4}	1.3×10^{-4}	2.6×10^{-6}	1.1×10^{-8}	4.3×10^{-11}
$x^{-\frac{3}{2}}$	$1 \le x < \infty$	3.0×10^{-4}	3.8×10^{-5}	1.4×10^{-6}	6.6×10^{-9}	2.6×10^{-11}
$\frac{1}{x^2+\sqrt{-1}}$	$-\infty < x < \infty$	2.3×10^{-3}	1.7×10^{-4}	2.6×10^{-6}	1.9×10^{-8}	8.1×10^{-11}
$e^{\frac{1}{z}}$	$\lvert z\rvert = 1$	7.6×10^{-5}	4.1×10^{-6}	4.8×10^{-8}	1.9×10^{-10}	7.3×10^{-13}

If the interval of integration is infinite, we compactify it to [0, 1] for the calculation. For example, if the interval is $(-\infty, \infty)$, the transformations

$$\phi_C : (0,1) \ni t \mapsto x = \tfrac{1}{C}\left(\tfrac{1}{1-t} - \tfrac{1}{t}\right) \in (-\infty, \infty),$$
$$\psi_C : (0,1) \ni t \mapsto x = \tfrac{1}{C}\left(e^{\frac{1}{1-t}} - e^{\frac{1}{t}}\right) \in (-\infty, \infty).$$

are used in [1]. In the above examples, the positive constant C is the default value in [1]. If $|f(x)| = O(x^{-2})$, the transformation by ϕ_C usually gives a better approximation than by ψ_C.

In Fig. 9 we show the change of integrand of $\displaystyle\int_{-\infty}^{\infty} \frac{dx}{x^2+1}$ under the compactification.

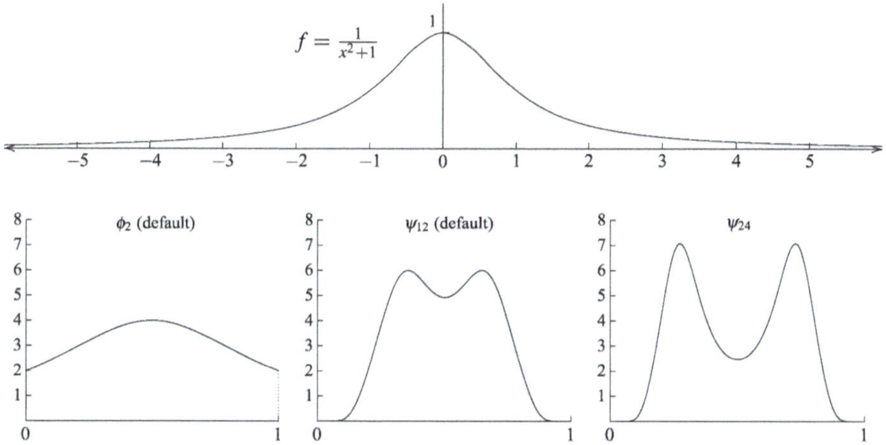

Fig. 9 Compactification

3.3 3D Graphs

Our original main purpose is to draw graphs of surfaces defined by $z = f(x, y)$ with mathematical functions $f(x, y)$. Using our method Fig. 1, we draw curves on a surface defined by the condition that x is constant or y is constant. It takes 10–30 s to get a required PDF file after a command in Risa/Asir if $f(x, y)$ is a simple rational function. We can use TikZ and XY-pic. In contrast to XY-pic the source text in TikZ is more readable, easy to be edited and has stronger abilities such that it supports coloring and filling region by a pattern but is takes a little longer time to be transformed into a PDF file. Hence our library [1] supports both of them (Fig. 10).

We give two examples $z = |\sin(x + y\sqrt{-1})|$ and $z = \frac{xy^2}{x^2+y^4}$.

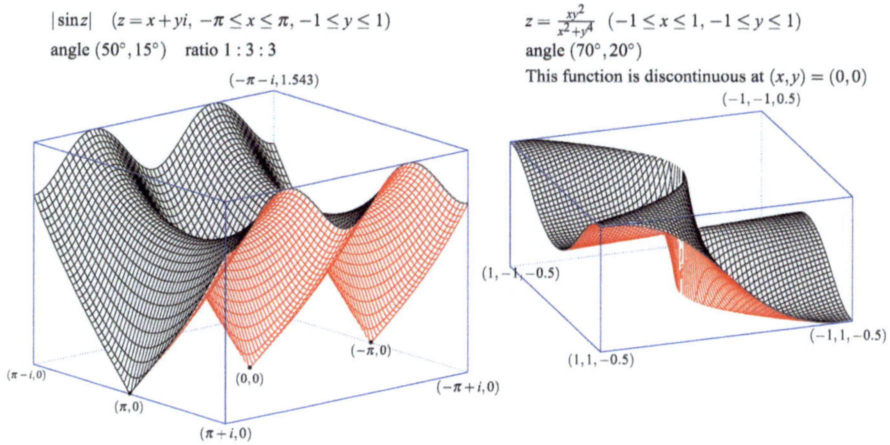

Fig. 10 3D graphs

References

1. T. Oshima, os_muldif.rr, a library for a computeralgebra Risa/Asir, 2008–2015, ftp://akagi.ms.u-tokyo.ac.jp/pub/math/muldif/
2. Risa/Asir, an open source general computer algebrasystem, http://www.math.kobe-u.ac.jp/Asir/asir.html
3. T. Oshima, *Fractional calculus of Weyl algebra and Fuchsiandifferential equations*, MSJ Memoirs. **28**, (Mathematical Society of Japan, 2012), http://projecteuclid.org/euclid.msjm/1413220558
4. E. Catmull, R. Rom, in *A class of local interpolating splines*, eds. by R.E. Barnhill, R.F. Reisenfeld, Computer Aided Geometric Design, (Academic Press, New York, 1974), pp. 317–326
5. P.J. Barry, R.N. Goldman, A recursive evaluation algorithm for a class of Catmull-Rom splines. SIGGRAPH Comput. Gr. **22**(4), 199–204 (1988)
6. C. Yuksel, S. Schaefer, J. Keyser, Parameterization and applications of Catmull-Rom curves. Comput.-Aided Des. **43**, 747–755 (2011)

Aesthetic Design with Log-Aesthetic Curves and Surfaces

Kenjiro T. Miura and R.U. Gobithaasan

Abstract Bézier, B-Spline and NURBS are de facto flexible curves developed for various design intent. However, these curves possess complex curvature function complicating the process of aesthetic shape design. In order to shorten the process, we introduce the fundmental equations of aesthetic curves and surfaces. This paper elaborates on the technicalities of Log-Aesthetic (LA) curves and surfaces along with its practical application for industrial design. It is anticipated that the emerging LA curves and surfaces have good prospects to be the standards for designing aesthetic shapes.

Keywords Fair curves and surfaces · Log-aesthetic curve and surface · General equations of aesthetic curves · Logarithmic curvature graph

1 Introduction

Fairness metric is a conventional term used to describe the quality of curves and surfaces. According to Farin, a planar curve is fair if it has continuous curvature and consists of few monotonic curvature pieces [1]. In the past, researchers formulated high-quality curves and surfaces using polynomial representation i.e. Bézier, B-spline and NURBS. The underlying principle was to obtain suitable functionals in order to minimize the oscillation of a curvature. Numerous principles are elaborated as chapters in a book edited by Sapidis entitled "Designing Fair Curves and Surfaces" [2].

A well defined Cesáro equation can be employed to produce high quality curves and surfaces. This equation expresses intrinsic properties of curves and surfaces

K.T. Miura (✉)
Shizuoka University, 3-5-1, Johoku, Naka-ku, Hamamatsu, Shizuoka, Japan
e-mail: tmkmiur@ipc.shizuoka.ac.jp

R.U. Gobithaasan
University Malaysia Terengganu, 21300 Kuala Terengganu, Terengganu, Malaysia
e-mail: gr@umt.edu.my

Y. Dobashi and H. Ochiai (eds.), *Mathematical Progress in Expressive Image Synthesis III*, Mathematics for Industry 24,
DOI 10.1007/978-981-10-1076-7_12

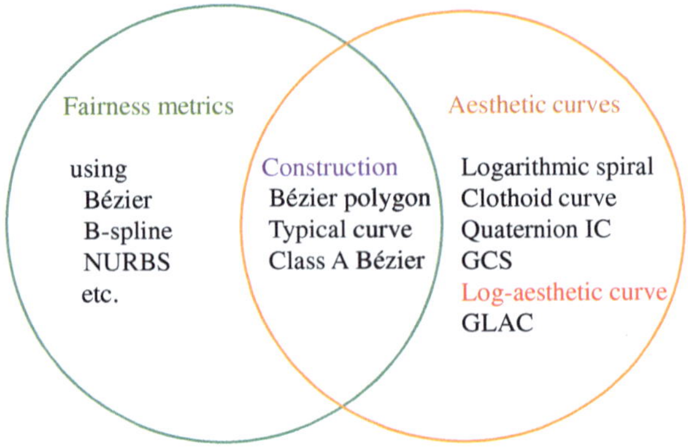

Fig. 1 Classifications of the researches on fair and aesthetic curves

without the presence of traditional polynomials. In this paper, we follow suit to distinguish the terms "fair" and "aesthetic". In brief, free-form curves and surfaces are constructed with specific formulations rather than polynomials or rationals to represent high quality shapes denoted as aesthetic curves and surfaces.

In recent years, researchers are perfecting the works on aesthetic curves contributing to exponential increase of findings and publications in this arena. However, research pertaining to the formulation of aesthetic surfaces are still at an early stage and offers bright opportunity for CAD community to contribute significantly. Figure 1 shows an Euler diagram depicting the classifications of fair and aesthetic curves. The members in the green circle comprises of the family which involves traditional curves. The orange circle encloses aesthetic curves including the log-aesthetic (LA) curves; which is one of the main focus of this paper. In this paper, we define aesthetic curves as non-polynomial functions used for aesthetic shape designs. The intersection of these circles represent the curves expressed by traditional formulations as well as specific methods of construction.

Recent advancement on the LA curve has been promising and it has now matured for industrial and graphical design practices. In 2009, Levien and Séquin [3] indicated that LA curves are the most promising curve for aesthetic design. Gobithaasan and Miura [4] formulated the generalized log-aesthetic curve (GLAC) in a standard form by representing the gradient of the logarithmic curvature graph (LCG) as a function of its arc length. They also reported that the LCG gradient of Generalized Cornu spiral [5] can be written as a linear function of the LCG gradient [6]. In 2012, Miura et al. reformulated the LA curve using variational principles to obtain minimized functionals for free-form surface design [7]. In the same year, Ziatdnov et al. [8] showed that some LA curves can be expressed by incomplete Gamma functions which shortens the computation time up to 10 times. Recently Meek et al. [9] proved that an unique solution exists for G^1 interpolation by using an LA curve segment when $\alpha < 0$. In 2015, Miura et al. [10] proposed another type of aesthetic curve called polar-aesthetic curve for scissors blade design.

2 Log-Aesthetic Curves

This section discusses about the details of LA curves and surfaces in a greater depth. We show that the formulation of LA curves has been perfected along with its fundamental properties. In the following sections we further illustrate that its applications for industrial design which proven to be promising. Albeit, the formulation and representation of aesthetic surfaces has been progressing and much efforts are anticipated for practical design. Next section is dedicated to dissect the foundation of LCG which led to the general equations of aesthetic curves.

2.1 Logarithmic Curvature Graphs

Harada's formulation of the Logarithmic Distribution Diagram of Curvature (LDDC) is based on a quantitative approach which involves tedious curvature radius and its arc length frequency calculation. They highlighted that an aesthetic curve has a linear LDDC plot. However, the length frequency of the curve can not be evaluated at certain positions on the curve or for arbitrary radius of curvature. Thus, it was not given much attention as a shape interrogation tool.

In 2003, Kanaya et al. [11] proposed the generation of Logarithmic Curvature Histogram, abbreviated LCH, to substitute LDDC and showed that for a given curve $\boldsymbol{C}(t) = (x(t), y(t))$, the derivative of the arc length s with respect to the logarithm of the radius of curvature $R = \log \rho$ is given by

$$\frac{ds}{dR} = \frac{(x'y'' - x''y')(x'^2 + y'^2)^{\frac{3}{2}}}{3(x'x'' + y'y'')(x'y'' - x''y') - (x'^2 + y'^2)(x'y''' - x'''y')} \tag{1}$$

where $'$ denotes the derivative with respect to parameter t. The quantitative LDDC approach is mathematically equivalent to Eq. (1) where horizontal and vertical coordinates represent R and $\log |ds/dR|$, respectively. Thus, Eq. (1) is sufficient to analytically define the LDDC plot. However, it does not provide concrete conditions in order to approximate LDDC with a straight line or it does indicate the slope of the approximated line to represent aesthetic shapes. Furthermore, for a curve whose shape is obtained from its image data, only discrete data are available and the partial arc length s_j must be a finite value to calculate the length frequency.

Hence algebraic manipulation was carried out to reformulate the LCH of Eq. (1). Since the LDDC graph is expressed by $\log |ds/d(\log \rho)|$ and both s and ρ are functions of the parameter t, we can further simplify as follows

$$\begin{aligned}\log\left|\frac{ds}{d(\log \rho)}\right| &= \log\left|\frac{\frac{ds}{dt}}{\frac{d(\log \rho)}{dt}}\right| = \log\left|\rho \frac{\frac{ds}{dt}}{\frac{d\rho}{dt}}\right| \\ &= \log \rho + \log s_d - \log\left|\frac{d\rho}{dt}\right|\end{aligned} \tag{2}$$

where $s_d = ds/dt$. Equation (2) is defined by the radius of curvature and its derivative. It describes the relationship between the radius of curvature and the derivative of the arc length more explicitly as compared to Eq. (1). This new analytical approach is denoted as Logarithmic Curvature Graph (LCG).

2.2 First and Second Fundamental Equations of Aesthetic Curves

In this section, we derive the equation of a curve that produces LCG strictly as a straight line. The curves obtained satisfying such a condition are defined as aesthetic curves and it is the fundamental equations of aesthetic curves [12].

If we assume that the LCG of a given curve is strictly expressed by a straight line, on the LCG of Eq. (2) there is a constant α and

$$\log\left|\rho\frac{ds}{d\rho}\right| = \alpha\log\rho + C \tag{3}$$

where C is a constant. By transforming Eq. (3), we obtain

$$\frac{1}{\rho^{\alpha-1}}\frac{ds}{d\rho} = e^C = C_0 \tag{4}$$

Hence,

$$\frac{ds}{d\rho} = C_0\rho^{\alpha-1} \tag{5}$$

If $\alpha \neq 0$,

$$s = \frac{C_0}{\alpha}\rho^\alpha + C_1 \tag{6}$$

In the above equation, C_1 is an integral constant. Therefore

$$\rho^\alpha = C_2 s + C_3 \tag{7}$$

where $C_2 = \alpha/C_0$ and $C_3 = -(C_1\alpha)/C_0$. Here we rename C_2 and C_3 to c_0 and c_1, respectively. Then

$$\rho^\alpha = c_0 s + c_1 \tag{8}$$

The above equation indicates that the αth power of the radius of curvature ρ is given by a linear function of the arc length s. The above equation is named the first fundamental equation of aesthetic curves [12].

For the case of $\alpha = 0$,

$$s = C_0 \log \rho + C_1 \tag{9}$$

Hence,

$$\rho = C_2 e^{C_3 s} \tag{10}$$

where $C_2 = e^{-C_1/C_0}$ and $C_3 = 1/C_0$. We rename C_2 and C_3 as c_0 and c_1, respectively. We get

$$\rho = c_0 e^{c_1 s} \tag{11}$$

The ρ is given by an exponential function of s. The above equation is named as the second fundamental equation of aesthetic curves [12].

It is known that logarithmic spiral and clothoid are regarded as high quality curves as explained in the next section. One of the principal characters of the logarithmic spiral is that its radius of curvature and arc length are proportional. Hence, the logarithmic spiral satisfies Eq. (8) and its α is equal to 1. Additionally, the main property of clothoid is that its radius of curvature is inversely proportion to its arc length. Thus, Eq. (8) is satisfied for the clothoid if α is given by -1.

In summary, the general equations of aesthetic curves expressed by Eq. (8) encompasses beautiful curves such as logarithmic spirals and clothoids. In fact Nielsen's spiral [13] is also expressed by Eq. (11). The curves which satisfies the first and second fundamental equations of aesthetic curves are denoted as Log-aesthetic (LA) curves, which was coined by Professor Carlo H. Séquin from University of California, Berkeley during the presentation of this work at the International CAD Conference & Exhibition 2007 at Honolulu, Hawaii.

2.3 *Parametric Expression of LA Curve*

In this section, we describe the parametric expressions of the fundamental equations for aesthetic curves as Eqs. (8) and (11). Let a curve $\boldsymbol{C}(s)$ satisfy Eq. (8). Then

$$\rho(s) = (c_0 s + c_1)^{\frac{1}{\alpha}} \tag{12}$$

Since s is the arc length, $|s_d| = 1$ (refer to, for example, [1]) and there exists $\theta(s)$ satisfying the following two equations:

$$\frac{dx}{ds} = \cos\theta, \quad \frac{dy}{ds} = \sin\theta \tag{13}$$

Since $\rho(s) = 1/(d\theta/ds)$,

$$\frac{d\theta}{ds} = (c_0 s + c_1)^{-\frac{1}{\alpha}} \tag{14}$$

Hence, if $\alpha \neq 1$,

$$\theta(s) = \frac{\alpha(c_0 s + c_1)^{\frac{\alpha-1}{\alpha}}}{(\alpha - 1)c_0} + c_2 \tag{15}$$

If the starting point of the curve is given by $\boldsymbol{P}_0 = \boldsymbol{C}(0)$, thus LA curve can be represented in a complex plane as follows

$$\boldsymbol{C}(s) = \boldsymbol{P}_0 + \exp(ic_2) \int_0^s \exp\left\{ i \frac{\alpha(c_0 u + c_1)^{\frac{\alpha-1}{\alpha}}}{(\alpha - 1)c_0} \right\} du \tag{16}$$

The above expression can be regarded as an extension of the clothoid curve where power of e in its definition is changed from 2 to $\alpha + 1$ and its LCG gradient can be specified to be equal to any value except for 0.

Similarly, the second fundamental equation of aesthetic curves are expressed by Eq. (11),

$$\frac{d\theta}{ds} = \frac{1}{c_0} e^{-c_1 s} \tag{17}$$

$$\theta = -\frac{1}{c_0 c_1} e^{-c_1 s} + c_2 \tag{18}$$

Therefore the curve is given by

$$\boldsymbol{C}(s) = \boldsymbol{P}_0 + e^{ic_2} \int_0^s e^{-\frac{i}{c_0 c_1} e^{-c_1 s}} ds \tag{19}$$

Figure 2 shows LA curves with various α values which are plotted with horizontal and vertical axes as real and imaginary values, respectively.

3 Log-Aesthetic Spline

In this section, we describe a method to simultaneously specify endpoints, tangent vectors and its curvatures (G^2 Hermite data) using an LA spline which consists three LA segments. The idea was obtained from Lan et al.'s work [14] where they solve the G^2 Hermite interpolation problem using triple clothoids. Miura et al. [15] used the shape parameter α as an additional parameter to make the end curvature as 0. Since α is related to impressions of the shape [16], hence α is fixed as a constant

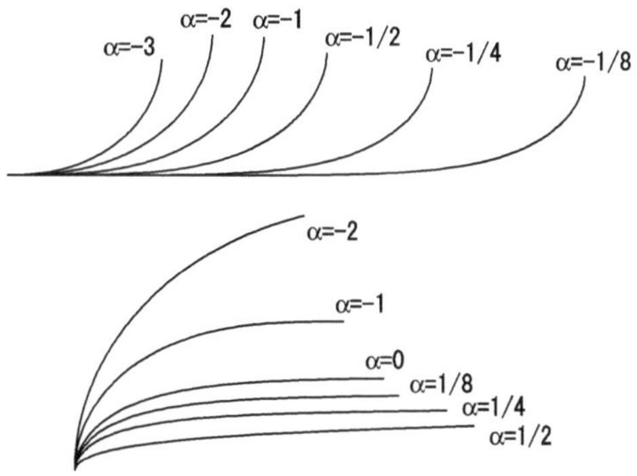

Fig. 2 LA curves whose LCG gradients are given by various α, *top* curves with negative α, *bottom* those with positive α and two negative values -2 and -1 for comparisons

to produce G^1 Hermite interpolation using a single LA curve segment and a C^3 continuous compound-rhythm LA curve is connected with two LA segments.

α is usually regarded as a shape parameter which can be controlled by designers. Thus, we do not use it to satisfy any geometric constraints to shape LA curves. Note that the degree of freedom (DoF) of the LA spline with three segments is similar to triple clothoids. It is a common practice to fix the value of α to design aesthetic shapes using LA curves. An LA spline consists of three LA curve segments with different α values which are joined with G^2 continuity [17].

3.1 Initial Value Estimation

We need initial values for parameters to define an LA spline. To obtain these initial values, we estimate curvatures at the two joints of the LA curve segments. We may use a Bézier curve of degree 5 for the estimation of the initial values to shape an LA spline. Let the total length of the Bézier curve as h. In general, Bézier curves are not uniquely determined by endpoints, tangent directions and its curvatures; these conditions do not necessarily yield a suitable curve for the initial value estimation. Hence we use an objective function which is modified to be independent from the total length h as proposed by Miura et al. [17]:

$$J_{LAC} = \frac{1}{h}\sum_{i=1}^{3}\int_{S_{i-1}h}^{S_i h}\sqrt{1+\alpha_i^2\rho^{2\alpha_i-2}\rho_s^2}ds \tag{20}$$

where $S_0 = 0$, $S_1 = 0.25$, $S_2 = 0.75$ and $S_3 = 1$. The above function is minimized to generate an appropriate initial Bézier curve. Figure 3a shows a Bézier curve of degree 5 and its initial control points in green and those after optimization in blue for $\alpha_i = -0.5$, $i = 1, 2, 3$.

Upon using the Bézier curve after optimization, an LA spline curve with three segments shown in red is determined. To be precise, the LA spline curve does not

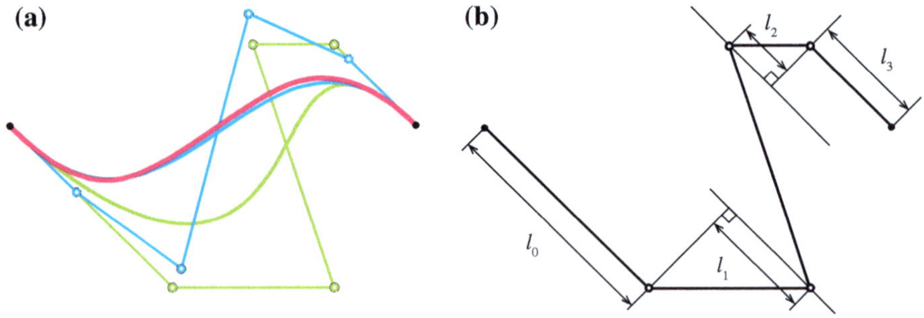

Fig. 3 Optimization of the approximation curve for initial values. **a** A triple LA curve. **b** Parameters for optimization

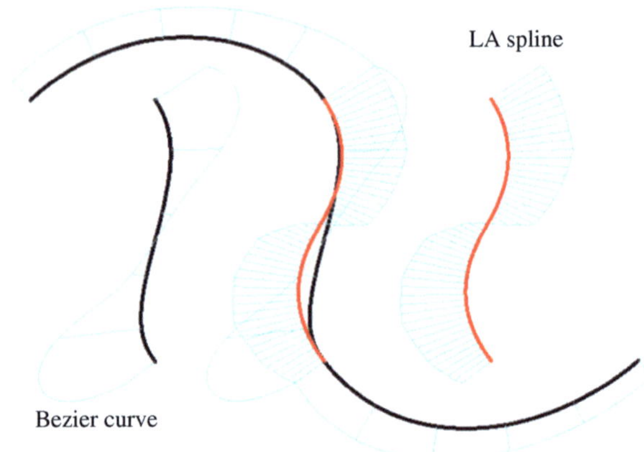

Fig. 4 A generation of G^2 continuous Bézier curve of degree five satisfying traditional G^2 Hermite data by a built-in command with and an LA spline in *red color* with the same G^2 Hermite data

strictly minimize the objective function in Eq. (20), but note that the shapes of the Bézier curve and LA spline are almost in a similar shape. In this example, parameters calculated from the input of Bézier curve are not appropriate, hence the numerical calculation diverges because the total length h becomes negative. If we use the Bézier curve after optimization, we may obtain these values without calculation failure and generate an LA spline successfully. Since an optimization process is necessary only for the initial value estimation, we propose a simpler method as shown in Fig. 3b. Let the lengths between the 1st and 2nd control points and the 5th and 6th control points be l_0 and l_3, respectively. Furthermore let parameters to determine the positions of the 3rd and 4th be l_1 and l_2, respectively. We change these parameters independently in the range of $0.05 \leq l_i/h \leq 0.5$ for $i = 0, 1, 2, 3$ by 0.05 where h is the total arc length of the input Bézier curve. We may now obtain parameter values easily which minimize the objective function in Eq. (20).

Figure 4 depicts Bézier curve of degree 5 (black) and an LA spline (red) generated using the proposed method. The Bézier curve is generated and deformed by built-in commands of a commercial CAD system to satisfy G^2 Hermite data posed at its endpoints. The LA spline is generated with similar G^2 Hermite data to achieve G^2

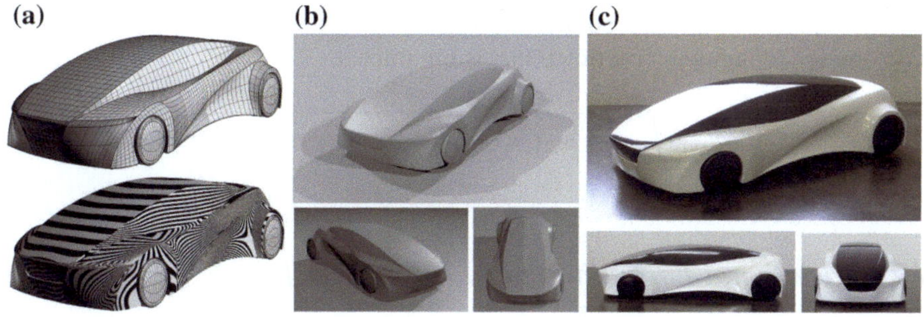

Fig. 5 A car model designed by using LA spline and its mock-up. **a** Iso-lines and zebra mapping. **b** Rendering. **c** Mock-up

continuity at the joints. These curves are also shown separately for visual clarity. The curvature plot of these curves is drawn in blue. It is visually clear that the curvature of the Bézier curve varies drastically in order to satisfy G^2 continuity at its endpoints whereas the LA spline joins gradually.

Figure 5 shows a practical design of a car using LA splines. Figure 5a shows iso-parametric lines of the free-form surface generated using LA splines and its corresponding zebra maps. Figure 5b depicts the CAD model with a special lighting condition and Fig. 5c are photos of its mock-up manufactured based on the CAD model. To note, the roof of the car is designed by a LA spline curve with three segments and its zebra maps indicates the surface is of high quality. Based on our experience on various aesthetic shape design, LA splines can be used to satisfy any G^2 Hermite data.

4 Variational Formulation of Aesthetic Shapes

In this section, we show on approximating aesthetic shapes with regard to the functional which LA curve minimizes [18]. Then, we extend the functional to formulate log-aesthetic surfaces [7].

4.1 Variational Formulation of LA Curves [18]

If we substitute ρ^α with σ in Eq. (8), the equation is then given by

$$\sigma = cs + d \tag{21}$$

The above equation indicates that LA curves are the representation of a straight line in the $s - \sigma$ plane where the horizontal and vertical axes are the arc length s and

$\sigma = \rho^\alpha$ respectively which connects two given points (s_1, β_1) and (s_2, β_2). In this case, the following objective functional is minimized.

$$J_{LAC} = \int_{s_1}^{s_2} \sqrt{1+\sigma_s^2 ds} = \int_{s_1}^{s_2} \sqrt{1+\alpha^2\rho^{2\alpha-2}\rho_s^2 ds} \tag{22}$$

4.2 *Variational Formulation of LA Surfaces [7]*

Here, we apply the variational principle to the surface formulation. We let the curvature of the curve κ and the arc length s correspond to the Gaussian curvature K and the surface area S, respectively. From Eq. (22), when $\alpha = -1$, $\kappa_s = -\rho_s/\rho^2$ and we obtain the following equation.

$$J_{LAC} = \int_{s_1}^{s_2} \sqrt{1+\kappa_s^2 ds} \tag{23}$$

By reparameterizing the above equation with $s = s(t)$, it becomes

$$J_{LAC} = \int_{t_1}^{t_2} \sqrt{x_t^2 + y_t^2 + \kappa_t^2 dt} = \int_{t_1}^{t_2} \sqrt{\lambda_C + \kappa_t^2 dt} \tag{24}$$

where $s_1 = s(t_1)$, $s_2 = s(t_2)$, and $\lambda_C = \sqrt{x_t^2 + y_t^2}$. Note that $ds/dt = \lambda_C$.

By extending Eq. (24) into the surface, we define the objective functional for the surface J_{LAS} as follows:

$$J_{LAS} = \int_{u_1}^{u_2} \int_{v_1}^{v2} \sqrt{\det(\boldsymbol{I}) + K_u^2 + K_v^2 dudv} \tag{25}$$

where $\boldsymbol{I}$ is a matrix expressed with the first fundamental form by

$$\boldsymbol{I} = \begin{bmatrix} E & F \\ F & G \end{bmatrix} \tag{26}$$

where $E = \boldsymbol{S}_u \cdot \boldsymbol{S}_u$, $F = \boldsymbol{S}_u \cdot \boldsymbol{S}_v$ and $G = \boldsymbol{S}_v \cdot \boldsymbol{S}_v$. Note that the area of the surface S is given by

$$S = \int_{u_1}^{u_2} \int_{v_1}^{v_2} \sqrt{\det(\boldsymbol{I})} dudv \tag{27}$$

We assume local parameterization (s, t) around a point on the surface so that the tangent vectors with respect to the parameters are orthogonal to each other, their directions are the same as the principal direction, and the norm of each tangent vector is equal to 1. With this parameterization, $\boldsymbol{I}$ becomes the 2×2 unit matrix. By performing integration around the point $\boldsymbol{S}(s_1, t_1)$, Eq. (25) can be rewritten as

$$\Delta J_{LAS} = \int_{s_1}^{s_1+\Delta s} \int_{t_1}^{t_1+\Delta t} \sqrt{1 + K_s^2 + K_t^2} dsdt \tag{28}$$

According to the general principle of variational principle, to minimize the following functional

$$J = \int_{s_1}^{s_2} \int_{t_1}^{t_2} g(K, K_s, K_t, s, t) dsdt \tag{29}$$

the following equation should be satisfied.

$$\frac{\partial g}{\partial K} - \frac{\partial}{\partial s}\frac{\partial g}{\partial K_s} - \frac{\partial}{\partial t}\frac{\partial g}{\partial K_t} = 0 \tag{30}$$

Note that $g = \sqrt{1 + K_s^2 + K_t^2}$ does not explicitly depend on K. Equation (30) yields

$$(1 + K_t^2)K_{ss} - 2K_sK_tK_{st} + (1 + K_s^2)K_{tt} = 0 \tag{31}$$

The above equation is called the minimal surface or Lagrange's equation and the surface $\boldsymbol{S}(s, t) = (s, t, K(s, t))$ is given by a minimal surface. Therefore, in a case where the Gaussian curvature on the boundary is specified, the Gaussian curvature should be given by a minimal surface interpolating the boundary values. The above discussion assumes local isometric parameterization whereby global isometric parameterization does not exist in general. It is not possible to deal with the case where the functional is defined globally as in Eq. (25). In such cases, some optimization technique should be adopted to minimize the functional to generate a desired surface.

According to Bernstein's theorem [19], if the boundary of the surface is located infinitely far, the minimal surface is given by a plane. Therefore, the Gaussian curvature is given by

$$K(s, t) = c_0 s + c_1 t + c_2 \tag{32}$$

where c_0, c_1, and c_2 are constants.

For further extension, we may use the mean curvature H instead of the Gaussian curvature K. In this paper we do not elaborate the effects of the power α. To take into account the effects of the power, we may use $\kappa_{max}^{\alpha}\kappa_{min}^{\beta}$ where κ_{max} and κ_{min} are the maximum and minimum normal curvatures, respectively. For example, an objective functional may be defined by

$$J_{LAS} = \int_{u_1}^{u_2} \int_{v_1}^{v_2} \sqrt{\det(\boldsymbol{I}) + \{(\kappa_{max}^{\alpha}\kappa_{min}^{\beta})_u\}^2 + \{(\kappa_{max}^{\alpha}\kappa_{max}^{\beta})_v\}^2} dudv \tag{33}$$

These extensions are proposed as topics for future research.

5 Conclusions

This paper reviews on fair curves and surfaces leading towards the formalization of aesthetic curves and surfaces. LA curves depict promising properties for practical applications and we hope it will be used as one of the standard curves for industrial and graphical design in the near future. Much effort for theoretical as well as practical researches are anticipated in order to define and utilize LA surface for various design feats.

Acknowledgments This work was supported in part by JSPS Grant-in-Aid for Scientific Research (B) Grant Number 25289021, JSPS Grant-in-Aid for Challenging Exploratory Research Grant Number 26630038, JST RISTEX Service Science, Solutions and Foundation Integrated Research Program and FRGS grant (FRGS: 59265) provided by University Malaysia Terengganu and Ministry of Education Malaysia.

References

1. G. Farin, *Curves and Surfaces for CAGD: A Practical Guide* (Morgan Kaufmann, San Francisco, 2002)
2. N.S. Sapidis (ed.), *Designing Fair Curves and Surfaces: Shape Quality in Geometric Modeling and Computer-Aided Design* (SIAM, Philadelphia, 1994)
3. R. Levien, C.H. Sequin, Interpolating splines: which is the fairest of them all? Comput.-Aided Des. Appl. **6**(1), 91–102 (2009)
4. R.U. Gobithaasan, K.T. Miura, Aesthetic spiral for design. Sains Malaysiana **40**(11), 1301–1305 (2011)
5. J.Md. Ali, R.M. Tookey, J.V. Ball, A.A. Ball, The generalized Cornu spiral and its application to span generation. J. Comput. Appl. Math. **102**(1), 37–47 (1999)
6. R.U. Gobithaasan, J.Md. Ali, K.T. Miura, The Elucidation of planar aesthetic curves, in *Proceedings of the 17th International Conference in Central Europe on Computer Graphics, Visualization and Computer Vision, WSCG2009*, pp. 138–188
7. K.T. Miura, R. Shirahata, S. Agari, S. Usuki, R.U. Gobithaasan, Variational formulation of the log-aesthetic surface and development of discrete durface filters. Comput.-Aided Des. Appl. **9**(6), 901–914 (2012)
8. R. Ziatdinov, N. Yoshida, T. Kim, Analytic parametric equations of log-aesthetic curves in terms of incomplete gamma functions. Comput. Aided Geom. Des. **29**(2), 129–140 (2012)
9. D.S. Meek, T. Saito, D.J. Walton, Y. Yoshida, Planar two-point g1 hermite interpolating log-aesthetic spirals. J. Comput. Appl. Math. **236**(17), 4485–4493 (2012)
10. K.T. Miura, R.U. Gobithaasan, S. Usuki, The polar-aesthetic curve and its applications to scissors design. Comput.-Aided Des. Appl. **12**(4), 431–438 (2014)
11. I. Kanaya, Y. Nakano, K. Sato, Simulated designer's eyes -classification of aesthetic surfaces, in *Proceedings of the VSMM* (2003), pp. 289–296
12. K.T. Miura, J. Sone, A. Yamashita, T. Kaneko, Derivation of a general formula of aesthetic curves, in *Proceedings of the Eighth International Conference on Humans and Computers (HC2005)* (2005), pp. 166–171
13. A. Gray, E. Abbena, S. Salamon, *Modern Differential Geometry of Curves and Surfaces with Mathematica* (Chapman & Hall, Boca Raton, 2006)
14. F. Lan, H. Tamai, H. Makino, Interpolation of arbitrary point sequence by triple clothoid curves. J. Jpn Soc. Precis. Eng. **76**(10), 1194–1199 (2010) (Japanese)

15. K.T. Miura, M. Yagi, Y. Kawata, M. Fujisawa, Input of aesthetic curve segments with inflection end points and generation of aesthetic curves with G2 continuity, in *Proceedings of the Graphics and CAD/Visual Computing Joint Symposium* (2007), pp. 297–302 (Japanese)
16. T. Harada, F. Yoshimoto, M. Moriyama, An aesthetic curve in the field of industrial design, in *Proceedings of the IEEE Symposium on Visual Language* (1999), pp. 38–47
17. K.T. Miura, D. Shibuya, R.U. Gobithaasan, S. Usuki, Designing log-aesthetic splines with G^2 continuity. Comput. Aided Des. Appl. **10**(6), 1021–1032 (2013)
18. K.T. Miura, S. Usuki, R.U. Gobithaasan, Variational formulation of the log-aesthetic curve, in *Proceedings of the 14th International Conference on Humans and Computers*, 9–10 March 2012, pp. 215–219
19. K. Kenmotsu, *Surfaces with Constant Mean Curvature* (American Mathematical Society, Providence, 2003)

Attractive Plane Curves in Differential Geometry

Jun-ichi Inoguchi

Abstract The purpose of chapter is to discuss plane curves from differential geometric point of view and applications of plane curves to computer aided designs. Plane curves are determined uniquely by curvatures up to Euclidean motions. Thus geometry of plane curves are formulated by the Euclidean motion group SE(2). From industrial point of view, other transformation groups are more appropriate for characterizing certain classes of plane curves. For instance, under equiaffine transformation group, conics are characterized as plane curves with constant equiaffine curvatures. Plane curves with monotonous curvature function have been paid much attention in industrial shape design and computer aided geometric design. In this chapter we study plane curves with monotonous curvature function, especially log-aesthetic curves, in terms of similarity transformation group.

Keywords Transformation group · Log-aesthetic curve · Similarity geometry · Similarity curvature

1 Introduction

Plane curves are fundamental objects in differential geometry. In industrial shape design or computer aided geometric design, curves of particular property have been used as a parts of figures. In particular, plane curves with monotonous curvature function have been paid much attention (see e.g., [30]).

For later use here we recall basic notations for Euclidean plane geometry. We denote by $\mathbb{R}^2$ the Cartesian plane with orthogonal coordinates (x, y). The *Euclidean distance function* $\mathrm{d} : \mathbb{R}^2 \times \mathbb{R}^2 \to \mathbb{R}$ is defined by

$$\mathrm{d}(\mathrm{P}, \mathrm{Q}) = \sqrt{(x-v)^2 + (y-w)^2}, \quad \mathrm{P} = (x, y), \mathrm{Q} = (v, w).$$

J. Inoguchi (✉)
Institute of Mathematics, University of Tsukuba, Tsukuba 305-8571, Japan
e-mail: inoguchi@math.tsukuba.ac.jp

Y. Dobashi and H. Ochiai (eds.), *Mathematical Progress in Expressive Image Synthesis III*, Mathematics for Industry 24,
DOI 10.1007/978-981-10-1076-7_13

Cartesian plane $\mathbb{R}^2$ equipped with Euclidean distance is called the *Euclidean plane* and denoted by $\mathbb{E}^2$.

A transformation f on $\mathbb{E}^2$ is said to be an *isometry* if it preserves the distance function d. Two figures X and Y in $\mathbb{E}^2$ are said to be *congruent each other* if there exits an isometry which sends X to Y.

It is a fundamental fact of Euclidean plane geometry, that every isometry on $\mathbb{E}^2$ has the form:

$$\mathbf{p} \longmapsto A\mathbf{p} + \mathbf{b},$$

where A is a real 2 by 2 orthogonal matrix and $\mathbf{b}$ is a vector.

The set of all isometries on $\mathbb{E}^2$ forms a Lie group with respect to composition. The resulting Lie group is called the *Euclidean group* and denoted by E(2). The Lie subgroup SE(2) of all orientation preserving isometries is called the *Euclidean motion group*. An element of SE(2) is called a *Euclidean motion*. Plane curves are determined by curvatures unique up to Euclidean motions.

In this chapter we replace SE(2) by other transformation groups and study attractive plane curves in terms of "curvatures" derived from transformation groups strictly larger than SE(2).

In particular, we discuss plane curves with *monotonous* curvature function via similarity transformation group.

2 Curves in Euclidean Plane Geometry

A plane curve $\mathbf{p}(t) = (x(t), y(t))$ is said to be *regular* if its derivative $\dot{\mathbf{p}}(t) = \mathrm{d}\mathbf{p}/\mathrm{d}t(t)$ never vanishes. Every regular plane curve can be reparametrized by the *arc length parameter* s. With respect to the arc length parameter s, the derivative $\mathbf{p}'(s) := \mathrm{d}\mathbf{p}/\mathrm{d}s(s)$ is a unit vector field along the curve. Moreover the vector field $\mathbf{T}(s) := \mathbf{p}'(s)$ and $\mathbf{N}(s) = R(\pi/2)\mathbf{T}(s)$ constitutes an orthonormal frame field $\{\mathbf{T}(s), \mathbf{N}(s)\}$ along the curve. Here $R(\theta)$ denotes the rotation matrix of rotation angle θ. The orthonormal frame field is regarded as an orthogonal matrix valued function $F(s) = (\mathbf{T}(s)\, \mathbf{N}(s))$. The matrix valued function $F(s)$ takes value in the *rotation group* SO(2) and called the *Frenet frame* of the curve. Here the rotation group SO(2) is a group of all rotation matrices, i.e., real 2 by 2 orthogonal matrices of determinant 1.

For an arc length parametrized curve $\mathbf{p}(s)$, its Frenet frame satisfies the *Frenet equation*:

$$F'(s) = F(s) \begin{pmatrix} 0 & -\kappa(s) \\ \kappa(s) & 0 \end{pmatrix}.$$

The function $\kappa(s)$ is called the *curvature* of $\mathbf{p}(s)$.

Theorem 1 (Fundamental theorem of plane curves) *Let* **p** *and* **q** *be regular plane curves defined on the same interval* I. *Assume that* **p** *and* **q** *have the same curvature. Then there exists a Euclidean motion mapping* **p** *to* **q**. *Hence* **p** *is congruent to* **q**.

For any function $\kappa = \kappa(s) : [0, L] \to \mathbb{R}$, there is an arc length parametrized curve $\mathbf{p} : [0, L] \to \mathbb{E}^2$ with arc length parameter s and curvature κ unique up to Euclidean motions. In fact $\mathbf{p}(s) = (x(s), y(s))$ is given explicitly by the following formula:

Theorem 2 (Representation formula)

$$x(s) = \int_0^s \cos\theta(s)\, \mathrm{d}s + x_0, \quad y(s) = \int_0^s \cos\theta(s)\, \mathrm{d}s + y_0,$$

$$\theta(s) := \int_0^s \kappa(s)\, ds + \theta_0,$$

where x_0, y_0 *and* θ_0 *are integral constants. The function* $\theta(s)$ *is called the turning angle function.*

Example 1 (*Logarithmic spirals*) A *logarithmic spiral* is a curve parametrized as

$$\mathbf{p}(t) = a(e^{bt} \cos t, e^{bt} \sin t).$$

Here a and b are positive constants (see Fig. 1). This logarithmic spiral has curvature $\kappa(s) = 1/(bs + a\sqrt{1+b^2})$ with $s = a\sqrt{1+b^2}(e^{bt} - 1)/b$. Thus logarithmic spirals are characterized as arc length parametrized curves whose *curvature radius* $\rho(s) = 1/\kappa(s)$ is a linear function of the arc length parameter. We shall give another characterization of logarithmic spirals in Sect. 4.

Example 2 (*Clothoids*) The *clothoid* (also called the *Cornu spiral*) is defined as

$$\mathbf{p}(s) = \left(\int_0^s \cos\frac{ks^2}{2}\, \mathrm{d}s, \int_0^s \sin\frac{ks^2}{2}\, \mathrm{d}s\right)$$

by using the *Fresnel integral*. Here k is a positive constant (Fig. 2). Clothoids have been used in *highway design* for many years. The clothoid has curvature $\kappa(s) = ks$. See also [18]. For some applications of Cornu spirals in CG, we refer to [3, 16, 17]. Bertails-Descoubes introduced the notion of super clothoid in [4]. See also [5].

By observing the explicit parametrization of the clothoids or the curvature function, the following two generalizations are proposed by differential geometers.

Example 3 (*Pseudo-spirals*) Gray suggested the following generalization of clothoids [9]:

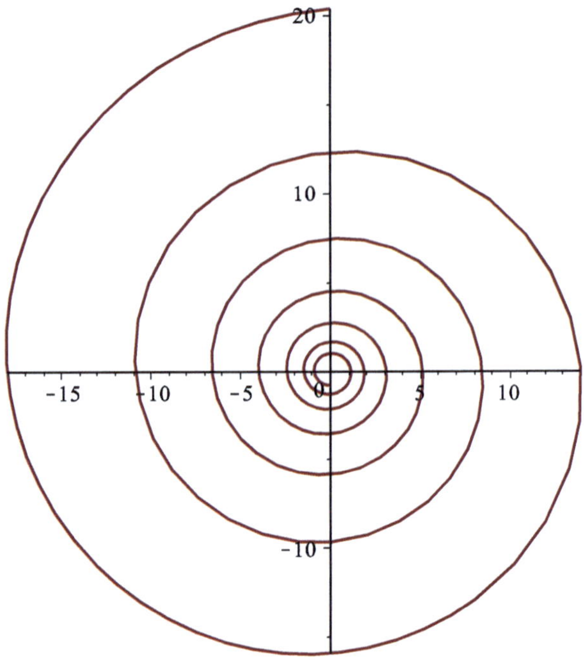

Fig. 1 Logarithmic spiral

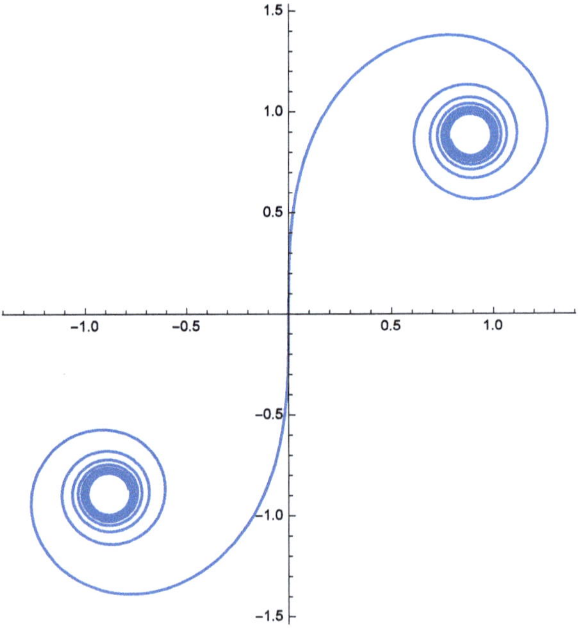

Fig. 2 Clothoid

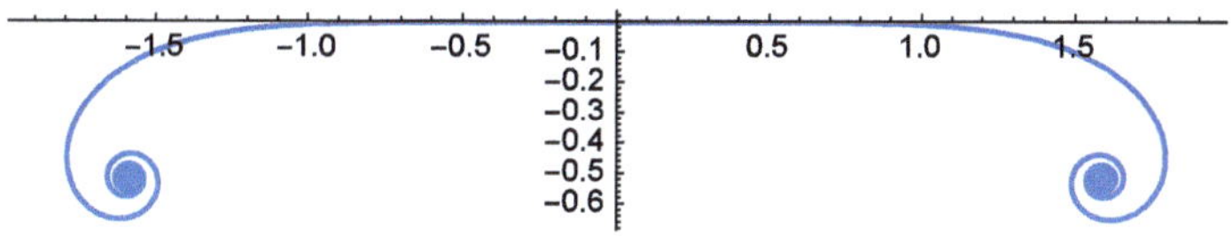

Fig. 3 Pseudo-spiral ($n = 2$)

$$x(t) = k \int_0^t \sin \frac{u^{n+1}}{n+1}\, du, \quad y(t) = k \int_0^t \cos \frac{u^{n+1}}{n+1}\, du.$$

Here k is a positive constant and n is a positive integer (Fig. 3).

The arc length parameter is $s(t) = kt$. The curvature is $\kappa(s) = -s^n/k^{n+1}$. This curve has explicit parametrization

$$x(t) = \frac{kt^{n+2}}{(n+1)(n+2)}\, {}_2F_1\left(\frac{1}{2} + \frac{1}{2(n+1)}, \frac{3}{2}; \frac{3}{2} + \frac{1}{2(n+1)}; -\frac{t^{2(n+1)}}{4(n+1)^2}\right),$$
$$y(t) = kt\, {}_2F_1\left(\frac{1}{2(n+1)}, \frac{1}{2}; 1 + \frac{1}{2(n+1)}; -\frac{t^{2(n+1)}}{4(n+1)^2}\right)$$

in terms of Gauss hypergeometric function ${}_2F_1$. An arc length parametrized curve is said to be a *pseudo-spiral* if its curvature function is given by $\kappa(s) = s^n/\alpha$. Here α is a positive constant and n is an integer. Pseudo-spirals have been considered by Savelov. Moreover Savelov pointed out that the involutes of pseudo-spirals are pseudospirals, too [26, p. 264, Eq. 6].

Remark 1 The hypergeometric function ${}_2F_1(a, b, c; t)$ is a solution to *Gauss hypergeometric differential equation*:

$$t(1-t)\frac{d^2 z}{dt^2} + \{c - (1 + a + b)t\}\frac{dz}{dt} - abz = 0, \tag{1}$$

where a, b and c are constants. The hypergeometric function ${}_2F_1(a, b; c; t)$ has the following expansion around 0:

$${}_2F_1(a, b; c; t) := \sum_{n=0}^{\infty} \frac{(a)_n (b)_n}{(c)_n\, n!} t^n, \quad |t| < 1,$$

where c is not a nonpositive integer. Here we used the so-called *Pochhammer symbol*

$$(a)_n := \begin{cases} a(a+1)(a+2)\ldots(a+n-1), & n \geq 1 \\ 1, & n = 0. \end{cases}$$

Example 4 (*Polynomial spirals*) Dillen [7] studied arc length parametrized curve whose curvature is a polynomial function of the arc length parameter (called *polynomial spirals*). See Fig. 4.

These two generalizations are derived from *Differential Geometric interests*. Here is another generalization suggested by Ali et al. [1] from computer design motivation.

Example 5 (*Span generation problem*) For any given data:

$$(\mathbf{p}_0, \mathbf{T}_0, \kappa_0, \theta_0), \quad (\mathbf{p}_1, \mathbf{T}_1, \kappa_1, \theta_1),$$

we look for arc length parametrized curve $\mathbf{p}(s)$, $(0 \leq s \leq L)$ such that

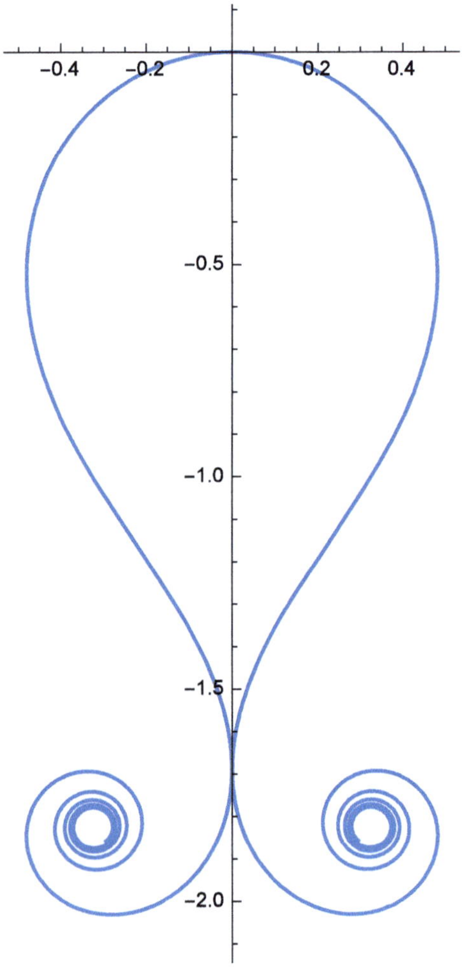

Fig. 4 Polynomial spiral with $\kappa(s) = s^2 - 2.19$

$$\mathbf{p}(0) = \mathbf{p}_0, \quad \dot{\mathbf{p}}(0) = \mathbf{T}_0, \quad \kappa(0) = \kappa_0, \quad \theta(0) = \theta_0,$$

$$\mathbf{p}(L) = \mathbf{p}_1, \quad \dot{\mathbf{p}}(L) = \mathbf{T}_1, \quad \kappa(L) = \kappa_1, \quad \theta(L) = \theta_1.$$

Ali et al. considered $\mathbf{p}(s)$ defined by the curvature function:

$$\kappa(s) = \frac{p + qs}{L + rs},$$

where p, q, r and L are constants (Fig. 5). An arc length parametrized curve $\mathbf{p}(s)$ determined by this rational curvature is called a *generalized Cornu spiral* in [1].

This family includes clothoids for $q \neq 0$ and $r = 0$. In addition, log-spirals are generalized Cornu spirals with $q = 0$ and $r \neq 0$.

Generalized Cornu spirals in the sense of [1] give solutions to the above span generation problem. In fact, if we choose

$$p := \kappa_0 L, \quad q := \kappa_1 - \kappa_0 + r\kappa_1, \quad r > -1$$

and integrating $\kappa(s)$ under the above initial conditions, we obtain an arc length parametrized curve $\mathbf{p}(s)$.

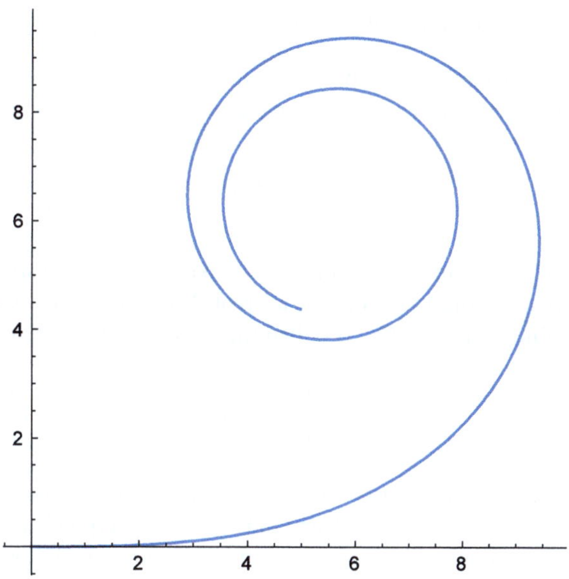

Fig. 5 Generalized Cornu spiral

3 Transformation Groups

From industrial point of view, we may replace the Euclidean motion group SE(2) by other transformation groups strictly larger than SE(2). Here is a mathematically rigorous definition of "plane geometry" in the spirit of Felix Klein (see [12]).

Definition 1 Let G be a Lie group acting transitively on $\mathbb{R}^2$. The pair $(G, \mathbb{R}^2)$ is called a *plane geometry* with *transformation group* G.

The Euclidean plane geometry is formulated as a plane geometry $(\mathrm{SE}(2), \mathbb{R}^2)$.

Example 6 (*Affine geometry*) Let us consider the Lie group A(2) of all *affine transformations* on $\mathbb{R}^2$. Affine transformations play fundamental roles in computer graphics. The plane geometry with transformation group A(2) is called the *affine plane geometry*. As is well known, every affine transformation on $\mathbb{R}^2$ has the form:

$$\mathbf{p} \longmapsto A\mathbf{p} + \mathbf{b},$$

where A is a real 2 by 2 nonsingular matrix and $\mathbf{b}$ is a vector. See [2, 24].

Example 7 (*Equiaffine geometry*) An affine transformation: $\mathbf{p} \longmapsto A\mathbf{p} + \mathbf{b}$ is said to be *equiaffine* if $|\det A| = 1$. The Lie group SA(2) of all equiaffine transformations is called the *equiaffine transformation group*. The plane geometry with equiaffine transformation group is called the *equiaffine plane geometry*.

Take a regular curve $\mathbf{p}(t)$. Assume that $\mathbf{p}(t)$ is *nondegenerate*, that is, $\det(\dot{\mathbf{p}}(t)\ddot{\mathbf{p}}(t)) \neq 0$. Then there exists a parameter u (called the *equiaffine parameter* or *unimodular parameter*) such that

$$F^{\mathrm{SA}}(u) = \left(\frac{d\mathbf{p}}{du}(u)\ \frac{d^2\mathbf{p}}{du^2}(u)\right)$$

has determinant 1. Namely F^{SA} takes value in the *special linear group* $\mathrm{SL}_2\mathbb{R}$. Here $\mathrm{SL}_2\mathbb{R}$ is the Lie group of all real 2 by 2 matrices with determinant 1. The matrix valued function F^{SA} is called the *equiaffine frame* and satisfies the *equiaffine Frenet equation*:

$$\frac{d}{du}F^{\mathrm{SA}}(u) = F^{\mathrm{SA}}(u)\begin{pmatrix} 0 & -\kappa^{\mathrm{SA}}(u) \\ 1 & 0 \end{pmatrix}.$$

The function $\kappa^{\mathrm{SA}}(u)$ is called the *equiaffine curvature* of $\mathbf{p}(u)$.

Analogus to Euclidean plane geometry, equiaffine parametrized plane curves are determined uniquely by equiaffine curvature up to orientation preserving equiaffine transformations.

Consider a parabola $\mathbf{p}(t) = (t, t^2/2)$. Although parabolas are fundamental examples of plane curves in Euclidean plane geometry, their curvature is not simple form. In fact, one can check that $\kappa(t) = (1 + t^2)^{-3/2}$. On the other hand in equiaffine

Fig. 6 Equiaffine clothoid

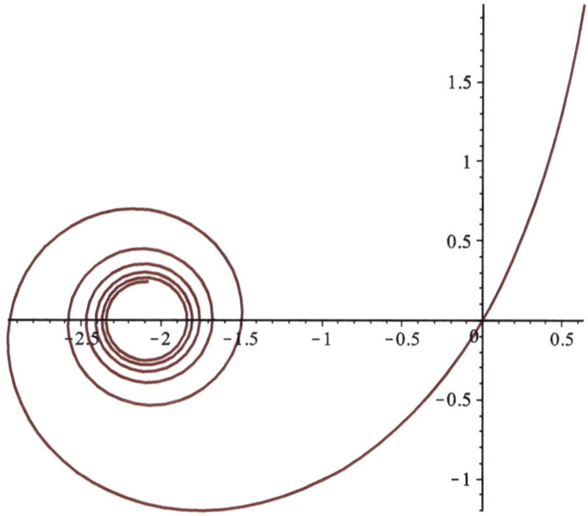

plane geometry, parabolas are characterized as equiaffine plane curves with *vanishing equiaffine curvature*. In addition, ellipses and hyperbolas are characterized as equiaffine curves with positive and negative constant equiaffine curvatures, respectively (see [23]).

Example 8 (*Equiaffine clothoid*) An *equiaffine clothoid* is an equiaffine plane curve whose equiaffine curvature is a linear function of the equiaffine parameter (Fig. 6). Equiaffine clothoid can be parametrized as

$$\mathbf{p}(u) = \sqrt{\pi}\left(\int_0^u \mathrm{Ai}(t)\,\mathrm{d}t,\ \int_0^u \mathrm{Bi}(t)\,\mathrm{d}t\right).$$

Here Ai and Bi are *Airy functions*.

4 Similarity Geometry

Hereafter, we concentrate our attention to plane curves under *similarity transformations*. The Lie group Sim(2) of all similarity transformations is generated by isometries and scalings.

The plane geometry under similarity transformations is called the *similarity plane geometry*. In similarity plane geometry, two geometric figures are said to be *equivalent* if they are related by similarity transformations each other.

Two circles C_1 and C_2 are *congruent*, that is, equivalent in *Euclidean plane geometry* if and only if they are related by isometries. Thus C_1 is congruent to C_2 if and only if the radii of two circles coincide. On the other hand, *any* two circles are equivalent in similarity plane geometry. Any two parabolas are mutually equivalent in similarity plane geometry.

As is well known, the logarithmic spiral is a *self-similar curve*. It is an interesting and natural question to characterize logarithmic spiral in terms of similarity geometry.

In similarity plane geometry, any regular plane curve with non-vanishing Euclidean curvature can be reparametrized by the turning angle function $\theta = \theta(s)$. Note that $\mathrm{d}\theta/\mathrm{d}s = \kappa(s)$. Analogues to Euclidean plane geometry, $\mathbf{p}(\theta)$ admits a matrix valued function $F^{\mathtt{Sim}}(\theta)$ which takes value in the matrix group

$$\mathrm{CO}^{+}(2) = \{Ar \mid r > 0, \quad A \in \mathrm{SO}(2)\}.$$

The *similarity frame* $F^{\mathtt{Sim}}(\theta)$ satisfies the similarity Frenet equation

$$\frac{\mathrm{d}}{\mathrm{d}\theta} F^{\mathtt{Sim}}(\theta) = F^{\mathtt{Sim}}(\theta) \begin{pmatrix} -S(\theta) & -1 \\ 1 & -S(\theta) \end{pmatrix}.$$

The function $S(\theta)$ is called the *similarity curvature* (cf. [6]). One can see that $\mathbf{p}(\theta)$ has vanishing similarity curvature if and only if $\mathbf{p}(\theta)$ is a circle. Regular plane curves with non-vanishing Euclidean curvature with non-zero constant similarity curvature are equivalent to logarithmic spirals. The similarity curvature $S(\theta)$ is related to Euclidean curvature $\kappa(s)$ by the formula:

$$S(\theta) = \frac{1}{\kappa(s)^2} \frac{\mathrm{d}\kappa}{\mathrm{d}s}(s).$$

Thus a Euclidean plane curve $\mathbf{p}(s)$ has *monotonous* Euclidean curvature if and only if its similarity curvature has *constant sign* as a similarity curve.

For more informations on similarity plane geometry, we refer to [13–15].

5 Log-Aesthetic Curves

Harada et al. [10, 11] considered *logarithmic distribution diagrams of curvature* (LDDC) for plane curves. They discovered that *attractive* plane curves drawn by car designers have approximately linear LDDC's. Note that logarithmic distribution diagrams of curvature are also called *logarithmic curvature graphs* (LCG).

The LDDC of an arc length parametrized curve $\mathbf{p}(s)$ is

$$(X(s), Y(s)) = (\log \rho(s), \log |\mathrm{d}s/\mathrm{d}\log \rho(s)|).$$

Miura [19] has given the following mathematical formulation of *log-aesthetic curve* (see also [21]). A log-aesthetic curve (of *slope* α) is an arc length parametrized curve $\mathbf{p}(s)$ whose LDDC is a straight line $Y = \alpha X + C$.

More explicitly, a log-aesthetic curve of slope α is an arc length parametrized curve $\mathbf{p}(s)$ whose curvature radius $\rho(s)$ is given by

$$\rho(s)^{\alpha} = as + b, \quad \alpha \neq 0$$

for some constants a and b. In case $\alpha = 0$, a log-aesthetic curve is defined by the equation:

$$\rho(s) = \exp(as + b)$$

for some constants a and b (see Fig. 7). Note that the class of log-aesthetic curves contains log-spirals ($\alpha = 1$), clothoids ($\alpha = -1$) and *Nielsen spirals* ($\alpha = 0$). The slope α is also called the *shape parameter*.

Some applications of log-aesthetic curves to car body design, we refer to a contribution [20] in this volume by Miura and Gobithaasan.

We notice that the Y-coordinate of the LDDC is rewritten as $Y = X - \log |S(\theta)|$. As a result, the slope $\alpha = \mathrm{d}Y/\mathrm{d}X$ is a *similarity invariant*, in fact, we have:

$$\alpha = 1 + \frac{1}{S^2}\frac{\mathrm{d}S}{\mathrm{d}\theta}. \tag{2}$$

Fig. 7 Log-aesthetic curve with $\alpha = -1/4$

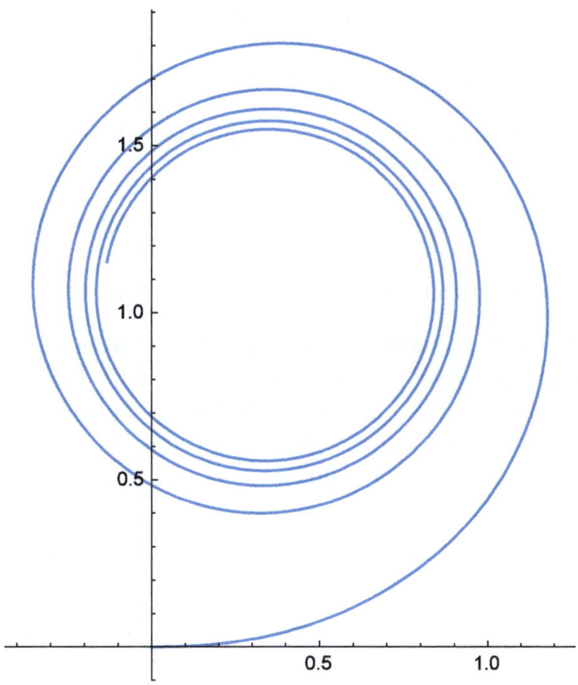

This fact motivates us to study log-aesthetic curves in terms of *similarity geometry*. Computing the similarity curvature of log-aesthetic curves we obtain the following fundamental result.

Theorem 3 *Log-aesthetic curves are characterized as plane curves whose reciprocal similarity curvature* $1/S(\theta)$ *is a linear function of* θ.

From (2), the similarity curvature $S(\theta)$ of log-aesthetic curves satisfies the ordinary differential equation:

$$\frac{\mathrm{d}S}{\mathrm{d}\theta} = -(1-\alpha)S^2.$$

Solving this ordinary differential equation, we have the general solution:

$$S(\theta) = -\frac{\lambda}{(\alpha-1)\lambda\theta+1} =: L(\alpha, \lambda, \theta), \quad \lambda \in \mathbb{R}.$$

Every solution $L(\alpha, \lambda, \theta)$ defines a log-aesthetic curve unique up to similarity transformations.

The curvature radius of a log-aesthetic curve is rewritten as ([27]):

$$\rho(\theta) = \begin{cases} \exp(\lambda\theta), & \alpha = 1 \\ \{(\alpha-1)\lambda\theta+1\}^{1/(\alpha-1)}, & \alpha \neq 1. \end{cases}$$

Here is an application of similarity geometric procedure for log-aesthetic curves.

Definition 2 For a plane curve $\mathbf{p}(\theta)$ in similarity plane geometry, its *similarity normal shift* $\hat{\mathbf{p}}$ of $\mathbf{p}$ is defined by

$$\hat{\mathbf{p}} := \mathbf{p}(\theta) + \mathbf{N}^{\mathrm{sim}}(\theta),$$

where $\mathbf{N}^{\mathrm{sim}}(\theta)$ is the similarity normal vector field.

One can see that the similarity normal shift $\hat{\mathbf{p}}$ coincides with the *evolute* of the original curve $\mathbf{p}$ in *Euclidean language*. Note that the original curve $\mathbf{p}$ is called the *involute* of $\hat{\mathbf{p}}$. The similarity normal shift of a log-aesthetic curve is also a log-aesthetic curve. For simplicity here we only treat the case $\alpha \neq 1, 2$. Then the similarity curvature of the similarity normal shift is

$$\hat{S}(\hat{\theta}) = L\left(\frac{1}{2-\alpha}, (2-\alpha)\lambda, \theta - \frac{\pi}{2}\right), \quad \hat{\theta} = \theta - \pi/2.$$

Hence the similarity normal shift of a log-aesthetic curves is also a log-aesthetic curve. In other words, similarity normal shifts preserve the class of log-aesthetic curves. This fact is a similarity geometric reformulation of a result due to Saito and Yoshida [28]. Thus the similarity normal shift is the *similarity geometric transformation*:

$$L(\alpha, \lambda, \theta) \longmapsto L\left(\frac{1}{2-\alpha}, (2-\alpha)\lambda, \theta - \frac{\pi}{2}\right)$$

between log-aesthetic curves.

Conversely, assume that $\hat{S}(\hat{\theta})$ satisfies the Riccati equation

$$\frac{\mathrm{d}\hat{S}}{\mathrm{d}\hat{\theta}} = \hat{S}(\hat{\theta})^2 - L\left(\frac{1}{2-\alpha}, (2-\alpha), \hat{\theta}\right)\hat{S}(\hat{\theta}).$$

Then the general solution of this ordinary differential equation is

$$\hat{S}(\hat{\theta}) = \frac{L(\alpha, \lambda, \hat{\theta})}{1 + \frac{1}{C\lambda\{(\alpha-1)\lambda\hat{\theta}+1\}^{\frac{1}{\alpha-1}}}}, \quad C \in \mathbb{R}. \tag{3}$$

We look for similarity plane curves determined by the similarity curvature $\hat{S}(\hat{\theta})$. To this end we recall here the notion of *ρ-shift generalized log-aesthetic curve* (ρ-shift GLAC) introduced by Gobithaasan and Miura [8].

The ρ-shift GLAC $\mathbf{p}_\nu$ with slope $\alpha \neq 1$ is a Euclidean plane curve determined by the curvature radius

$$\rho_\nu(\theta) = \{(\alpha-1)\lambda\theta + 1\}^{\frac{1}{\alpha-1}} + \nu, \quad \nu \in \mathbb{R}.$$

The similarity curvature S_ν of a ρ-shift GLAC $\mathbf{p}_\nu$ is

$$S_\nu(\theta) = \frac{L(\alpha, \lambda, \theta)}{1 + \frac{\nu}{\{(\alpha-1)\lambda\theta+1\}^{\frac{1}{\alpha-1}}}}. \tag{4}$$

Choose $\nu = 1/(C\lambda)$ in the formula (4) and comparing it with (3), we obtain

Theorem 4 ([25]) *ρ-shift* GLACs *are involutes of log-aesthetic curves.*

In addition Miura et al. obtained a similarity geometric reformulation/interpretation of κ-shift generalized aesthetic curves [22].

These facts actually show that similarity geometry is really useful for the study of aesthetic curves. More examples of log-aesthetic curves can be obtained from certain Riccati equation for similarity curvature.

We close this chapter with exhibiting Euclidean plane curves whose similarity curvature satisfies the Riccati equation of the form:

$$\frac{\mathrm{d}S}{\mathrm{d}\theta} = S^2 + \frac{a_1\theta + a_0}{b_2\theta^2 + b_1\theta + b_0}S + \frac{d}{c_2\theta^2 + c_1\theta + c_0}.$$

Example 9 (*Parabola*) The similarity curvature of a parabola $(t, t^2/2)$ is $S(\theta) = -3\tan\theta$ with $\theta = \tan^{-1} t$. This similarity curvature is a solution to

$$\frac{dS}{d\theta} = S^2 - 3.$$

Example 10 (*Superspiral*) A *superspiral* in the sense of Ziatdinov [29] is an arc length parametrized curve whose curvature radius $\rho(s)$ is given by

$$\rho(s) = {}_2F_1(a, b; c; -\theta(s)),$$

where $a > 0$ and $0 < b < c$. For $\theta \geq 0$, the curvature of a superspiral is monotonous.

One can check that the similarity curvature of a superspiral satisfies the Riccati equation:

$$\frac{dS}{d\theta} = S^2 + \frac{(a+b+1)\theta + c}{\theta(1+\theta)} S - \frac{ab}{\theta(1+\theta)}.$$

Discussions in this chapter motivate us to study attractive curves used in computer graphics or computer aided geometric design in terms of plane geometry under transformation groups *strictly larger* than Euclidean motion groups.

Acknowledgments The author would like to thank the organizers for inviting him to the workshop MEIS2015. The author would also like to thank Professor Kenjiro T. Miura, Mr. Masayuki Sato, Mr. Yasuhiro Shimizu and Professor Rushan Ziatdinov for their useful comments.

This work is partially supported by KAKENHI 15K04834.

References

1. J.M. Ali, R.M. Tookey, J.V. Ball, A.A. Ball, The generalised Cornu spiral and its application to span generation. J. Comput. Appl. Math. **102**(1), 37–47 (1999)
2. K. Anjyo, H. Ochiai, *Mathematical Basics of Motion and Deformation in Computer Graphics* (Morgan & Claypool, San Rafael, 2014)
3. I. Baran, J. Lehtinen, J. Popović, Sketching clothoid splines using shortest paths. Comp. Graph. Forum **29**, 655–664 (2010)
4. F. Bertails-Descoubes, Super-clothoids. Comput. Graph. Forum **31**(2pt2), 509–518 (2012) (Proceedings of the Eurographics' 12)
5. F. Bertails-Descoubes, Geometry and mechanics of fibers: some numerical models. in *Mathematical Progress in Expressive Image Synthesis III*, ed. by Y. Dobashi, H. Ochiai (Springer, Singapore, 2016). doi:10.1007/978-981-10-1076-7_1
6. E. Cartan, La Méthode de Repèere Mobile, la Thèeorie des Groupes Continus, et les Espaces Gééneralisés, In: Exposés de Géométrie V, Hermann, Paris (1935)
7. F. Dillen, The classification of hypersurfaces of a Euclidean space with parallel higher order fundamental form. Math. Z. **203**, 635–643 (1990)
8. R.U. Gobithaasan, K.T. Miura, Aesthetic spiral for design. Sains Malaysiana **40**, 1301–1305 (2011)
9. A. Gray, E. Abbena, S. Salamon, *Modern Differential Geometry of Curves and Surfaces with Mathematica*, 3rd edn. (CRC Press, Boca Raton, 2006)
10. T. Harada, Study of quantitative analysis of the characteristics of a curve. Forma **12**(1), 55–63 (1997)
11. T. Harada, N. Mori, K. Sugiyama, Study of quantitative analysis of curve's character (in Japanese). Bull. JSSD **40**, 9–16 (1994)

12. J. Inoguchi, *Introduction to Klein geometries (Kikagaku Iro Iro, in Japanese)* (Nippon Hyoronsha, Tokyo, 2007)
13. J. Inoguchi, *Plane Curves and Solitons (Kyokusen to Soliton, in Japanese)* (Asakura Publishing, Tokyo, 2010)
14. J. Inoguchi, *Can We Solve Riccati equations ? (Riccati no Himitsu, in Japanese)* (Nippon Hyoronsha, Tokyo, 2010)
15. K. Kajiwara, T. Kuroda, N. Matsuura, Isogonal deformation of discrete plane curves and discrete Burgers hierarchy. Pac. J. Math. Ind. **8**, 3 (2016)
16. B. Kimia, I. Frankel, A. Popescu, Euler spiral for shape completion. Int. J. Comput. Vision **54**, 157–180 (2003)
17. J. McCrae, K. Singh, Sketching piecewise clothoid curves. Comput. Graph. **33**, 452–461 (2009)
18. D.S. Meek, D.J. Walton, The use of Cornu spirals in drawing planar curves of controlled curvature. J. Comput. Appl. Math. **25**, 69–78 (1989)
19. K.T. Miura, A general equation of aesthetic curves and its self-affinity. Comput.-Aided Des. Appl. **3**, 457–464 (2006)
20. K.T. Miura, R.U. Gobithaasan, Aesthetic design with log-aesthetic curves and surfaces. in *Mathematical Progress in Expressive Image Synthesis III*, ed. by Y. Dobashi, H. Ochiai (Springer, Singapore, 2016). doi:10.1007/978-981-10-1076-7_12
21. K.T. Miura, J. Sone, A. Yamashita, T. Kaneko, Derivation of a general formula of aesthetic curves, in *Proceedings of the 8th International Conference on Humans and Computers (HC2005)* (Aizu-Wakamutsu, Japan, 2005), pp. 166–171
22. K.T. Miura, R.U. Gobithaasan, S. Suzuki, S. Usuki, Reformulation of generalized logaesthetic curves with Bernoulli equations. Comput.-Aided Des. Appl. **13**(2), 265–269 (2016)
23. K. Nomizu, T. Sasaki, *Affine Differential Geometry. Geometry of Affine Immersions* (Cambridge University Press, Cambridge, 1994)
24. H. Ochiai, K. Anjyo, Mathematical formulation of motion and deformation and its applications, *Mathematical Progress in Expressive Image Synthesis I*, vol. 4, Mathematics for Industry (Springer, Tokyo, 2014), pp. 123–129
25. M. Sato, Y. Shimizu, Log-aesthetic curves and Riccati equations from the viewpoint of similarity geometry. JSIAM Lett. **7**, 21–24 (2015)
26. A.A. Savelov, in *Plane Curves: Systematics, Properties, Applications (in Russian)*, ed. by A.P. Norden (Gosudarstv. Izdat. Fiz.-Mat. Lit, Moscow, 1960)
27. N. Yoshida, T. Saito, Interactive aesthetic curve segments. Vis. Comput. **22**, 9–11 (2006) (Proceedings of the Pacific Graphics)
28. N. Yoshida, T. Saito, On the evolutes of log-aesthetic planar curves (in Japanese). Research Report A of College of Industrial Technology. Nihon University **44**, 1–5 (2011)
29. R. Ziatdinov, Family of superspirals with completely monotonic curvature given in terms of Gauss hypergeometric function. Comput. Aided Geom. Des. **29**, 510–518 (2012)
30. R. Ziatdinov, R.I. Nabiyev, K.T. Miura, MC-curves and aesthetic measurements for pseudospiral curve segments. Math. Des. Tech. Aesthet. **1**(1), 6–17 (2013)

dNLS Flow on Discrete Space Curves

Sampei Hirose, Jun-ichi Inoguchi, Kenji Kajiwara, Nozomu Matsuura and Yasuhiro Ohta

Abstract The local induction equation, or the binormal flow on space curves is a well-known model of deformation of space curves as it describes the dynamics of vortex filaments, and the complex curvature is governed by the nonlinear Schrödinger equation (NLS). In this paper, we present its discrete analogue, namely, a model of deformation of discrete space curves by the discrete nonlinear Schrödinger equation (dNLS). We also present explicit formulas for both NLS and dNLS flows in terms of the τ function of the 2-component KP hierarchy.

Keywords Discrete space curve · Vortex filamant · Local induction equation · Discrete nonlinear Schrödinger equation · Soliton · τ function

S. Hirose
Center for Promotion of Educational Innovation, Shibaura Institute of Technology, 307 Fukasaku, Minuma-ku, Saitama 337-8570, Japan
e-mail: hirose3@shibaura-it.ac.jp

J. Inoguchi
Institute of Mathematics, University of Tsukuba, Tsukuba 305-8571, Japan
e-mail: inoguchi@math.tsukuba.ac.jp

K. Kajiwara (✉)
Institute of Mathematics for Industry, 744 Motooka, Nishi-ku, Fukuoka 819-0395, Japan
e-mail: kaji@imi.kyushu-u.ac.jp

N. Matsuura
Department of Applied Mathematics, Fukuoka University, Nanakuma 8-19-1, Fukuoka 814-0180, Japan
e-mail: nozomu@fukuoka-u.ac.jp

Y. Ohta
Department of Mathematics, Kobe University, Rokko, Kobe 657-8501, Japan
e-mail: ohta@math.sci.kobe-u.ac.jp

Y. Dobashi and H. Ochiai (eds.), *Mathematical Progress in Expressive Image Synthesis III*, Mathematics for Industry 24,
DOI 10.1007/978-981-10-1076-7_14

1 Introduction

The local induction equation (LIE)

$$\frac{\partial \gamma}{\partial t} = \frac{\partial \gamma}{\partial x} \times \frac{\partial^2 \gamma}{\partial x^2}, \tag{1}$$

is one of the most important models of deformation of space curves, where $\gamma(x, t) \in \mathbb{R}^3$ is a smooth space curve parametrized by the arc-length x and t is a deformation parameter [6, 13, 16]. In a physical setting, it describes the dynamics of vortex filaments driven by the self-induction in the inviscid fluid under the local induction approximation [6].

It is well-known that if γ obeys LIE, then the curvature and the torsion, or equivalently, the complex curvature of γ solves the *nonlinear Schrödinger equation* (NLS) which is one of the most typical equations in the integrable systems. To show this, we use the Frenet frame $\Phi = \Phi(x, t) = [T(x, t), N(x, t), B(x, t)] \in \mathrm{SO}(3)$, where T, N, B are the tangent, the normal, and the binormal vectors defined by

$$T = \gamma', \quad N = \frac{\gamma''}{|\gamma''|}, \quad B = T \times N, \quad ' = \frac{\partial}{\partial x}, \tag{2}$$

respectively. Note that it follows that $|T| = |\gamma'| = 1$ since x is the arc-length. Then we have the *Frenet–Serret formula*

$$\frac{\partial \Phi}{\partial x} = \Phi L, \quad L = \begin{bmatrix} 0 & -\kappa & 0 \\ \kappa & 0 & -\lambda \\ 0 & \lambda & 0 \end{bmatrix}, \tag{3}$$

where $\kappa = |\gamma''|$ and $\lambda = -\langle B', N \rangle$ are the curvature and the torsion, respectively. In this setting, LIE (1) is expressed as the deformation by the binormal flow

$$\frac{\partial \gamma}{\partial t} = \kappa B, \tag{4}$$

and the corresponding deformation equation of the Frenet frame is given by

$$\frac{\partial \Phi}{\partial t} = \Phi M, \quad M = \begin{bmatrix} 0 & \kappa\lambda & -\kappa' \\ -\kappa\lambda & 0 & -\frac{\kappa''}{\kappa} + \lambda^2 \\ \kappa' & \frac{\kappa''}{\kappa} - \lambda^2 & 0 \end{bmatrix}. \tag{5}$$

The compatibility condition of the system of linear partial differential equations (3) and (5)

$$\frac{\partial L}{\partial t} - \frac{\partial M}{\partial x} = LM - ML, \tag{6}$$

yields

$$\frac{\partial \kappa}{\partial t} = -2\frac{\partial \kappa}{\partial x}\lambda - \kappa\frac{\partial \lambda}{\partial x}, \quad \frac{\partial \lambda}{\partial t} = \frac{\partial}{\partial x}\left(\frac{\kappa''}{\kappa} + \frac{\kappa^2}{2} - \lambda^2\right). \tag{7}$$

Introducing the *complex curvature* $u = u(x, t) \in \mathbb{C}$ by the *Hasimoto transformation* [6]

$$u = \kappa e^{\sqrt{-1}\Lambda}, \quad \Lambda = \int^x \lambda\, dx, \tag{8}$$

we see that u satisfies NLS

$$\sqrt{-1}\frac{\partial u}{\partial t} + \frac{\partial^2 u}{\partial x^2} + \frac{1}{2}|u|^2 u = 0. \tag{9}$$

Also, one can show that this deformation is isoperimetric, namely $|\gamma'| = 1$ for all t.

Discretization of curves and their deformations preserving underlying integrable structures is an important problem in the discrete differential geometry. For example, the isoperimetric deformation of plane discrete curves described by the discrete mKdV equation (dmKdV) has been studied in [10, 12, 14]. For discrete space curves, the deformations by the discrete sine-Gordon equation (dsG) and dmKdV has been studied in [4, 11, 12], and the deformation by dNLS is formulated in [9, 17].

The dsG and dmKdV describe torsion-preserving isoperimetric and equidistant deformation of the space discrete curves with constant torsion. However, formulation of discrete deformation of space discrete curves with varying torsion is a difficult problem. The only example so far is presented by Hoffmann [8, 9], where he has claimed that composition of certain two isoperimetric equidistant deformations can be regarded as a discrete analogue of LIE. Also, it was used for numerical simulation of fluid flow [17, 19]. This formulation uses quaternions and its geometric meaning is clear, but description of the deformation parameters in terms of the complex curvature, thus the relation to dNLS are rather indirect.

In this paper, we present a formulation of the dNLS flow on discrete space curves from different approach; the deformation of curves is expressed in terms of the discrete Frenet frame with the coefficients given by the curvature and torsion of the curves explicitly. In this approach, dNLS arises as the equation governing the complex curvature of curves, which is the same as the case of smooth curves. Based on this formulation, we present explicit formulas for the NLS flow for smooth curves and the dNLS flow to discrete curves in terms of τ functions of the two-component KP hierarchy by applying the theory of integrable systems. We expect that our dNLS flow can be an alternative to Hoffmann's formulation when it is used to simulate the dynamics of fluids. Also, explicit expression of the scheme and exact solutions may promote further development of theoretical studies of discrete dynamics of discrete curves from both mathematical and physical point of view.

2 dNLS Flow on Discrete Space Curves

Let $\gamma_n \in \mathbb{R}^3$ be a discrete space curve with

$$|\gamma_{n+1} - \gamma_n| = \varepsilon, \tag{10}$$

where ε is a constant. We introduce the *discrete Frenet frame* $\Phi_n = [T_n, N_n, B_n] \in$ SO(3) by

$$T_n = \frac{\gamma_{n+1} - \gamma_n}{\varepsilon}, \quad B_n = \frac{T_{n-1} \times T_n}{|T_{n-1} \times T_n|}, \quad N_n = B_n \times T_n. \tag{11}$$

Then it follows that the discrete Frenet frame satisfies the *discrete Frenet–Serret formula*

$$\Phi_{n+1} = \Phi_n L_n, \quad L_n = R_1(-\nu_{n+1}) R_3(\kappa_{n+1}), \tag{12}$$

where

$$R_1(\theta) = \begin{bmatrix} 1 & 0 & 0 \\ 0 & \cos\theta & -\sin\theta \\ 0 & \sin\theta & \cos\theta \end{bmatrix}, \quad R_3(\theta) = \begin{bmatrix} \cos\theta & -\sin\theta & 0 \\ \sin\theta & \cos\theta & 0 \\ 0 & 0 & 1 \end{bmatrix}, \tag{13}$$

and ν_n, κ_n are defined by

$$\begin{gathered} \langle T_{n-1}, T_n \rangle = \cos\kappa_n, \quad \langle B_n, B_{n-1} \rangle = \cos\nu_n, \quad \langle B_n, N_{n-1} \rangle = \sin\nu_n, \\ -\pi \le \nu_n < \pi, \quad 0 < \kappa_n < \pi. \end{gathered} \tag{14}$$

In order to formulate a "good" discrete deformation (discretization of time), we resort to the theory of discrete integrable systems to preserve integrable nature of the NLS flow (4). As a discrete analogue of NLS (9), we consider

$$\begin{gathered} \left(\sqrt{-1}\tfrac{\varepsilon^2}{\delta} - 1\right) u_n^{m+1} - \left(\sqrt{-1}\tfrac{\varepsilon^2}{\delta} + 1\right) u_n^m + (u_{n+1}^m + u_{n-1}^{m+1})(1 + \varepsilon^2 |u_n^m|^2)\Gamma_n^m = 0, \\ \frac{\Gamma_{n+1}^m}{\Gamma_n^m} = \frac{1 + \varepsilon^2 |u_n^m|^2}{1 + \varepsilon^2 |u_n^{m+1}|^2}, \end{gathered} \tag{15}$$

which we refer to as the *discrete nonlinear Schrödinger equation* (dNLS) [1, 2, 18]. Here, $u_n^m \in \mathbb{C}$, $\Gamma_n^m \in \mathbb{R}$, n is the space discrete variable which corresponds to the label of vertices of discrete curves, m is the discrete variable corresponding to the step of deformation, ε and δ are constants which are the lattice intervals of n and m, respectively. Moreover, u_n^m is the complex discrete curvature defined by

$$u_n^m = \frac{1}{\varepsilon} \tan\frac{\kappa_n^m}{2} e^{\sqrt{-1}\Lambda_n^m}, \quad \Lambda_n^m - \Lambda_{n-1}^m = -\nu_n^m, \tag{16}$$

We impose the boundary condition as

$$u_n^m \to 0 \ (n \to \pm\infty), \quad \Gamma_n^m \to \Gamma_{\pm\infty} \ (\text{const.}) \ (n \to \pm\infty). \tag{17}$$

Then one of the main statements of this paper is given as follows:

Theorem 1 (dNLS flow) *For a fixed m, let $\gamma_n^m \in \mathbb{R}^3$ be a discrete space curve satisfying*

$$|\gamma_{n+1}^m - \gamma_n^m| = \varepsilon, \tag{18}$$

and $\Phi_n^m = [T_n^m, N_n^m, B_n^m] \in \mathrm{SO}(3)$ be the discrete Frenet frame defined in (11) satisfying the discrete Frenet–Serret formula

$$\Phi_{n+1}^m = \Phi_n^m L_n^m, \quad L_n^m = R_1(-\nu_{n+1}^m) R_3(\kappa_{n+1}^m). \tag{19}$$

Let u_n^m be a complex discrete curvature of γ_n^m. We determine u_n^{m+1} by dNLS (15) under the boundary condition (17) and put $u_n^{m+1} = \frac{1}{\varepsilon} \tan \frac{\kappa_n^{m+1}}{2} e^{\sqrt{-1}\Lambda_n^{m+1}}$. We define a new curve $\gamma_n^{m+1} \in \mathbb{R}^3$ by

$$\frac{\gamma_n^{m+1} - \gamma_n^m}{\delta} = \frac{2}{\varepsilon^3}(P_n^m T_n^m + Q_n^m N_n^m + R_n^m B_n^m), \tag{20}$$

$$\begin{aligned} P_n^m &= \delta\left(-1 + \frac{\Gamma_n^m}{\cos^2 \frac{\kappa_n^m}{2}}\right), \\ Q_n^m &= \delta\left[\tan \frac{\kappa_n^m}{2} - \tan \frac{\kappa_{n-1}^{m+1}}{2} \cos(\Lambda_{n-1}^{m+1} - \Lambda_n^m) \frac{\Gamma_n^m}{\cos^2 \frac{\kappa_n^m}{2}}\right], \\ R_n^m &= \varepsilon^2 \tan \frac{\kappa_n^m}{2} - \delta \tan \frac{\kappa_{n-1}^{m+1}}{2} \sin(\Lambda_{n-1}^{m+1} - \Lambda_n^m) \frac{\Gamma_n^m}{\cos^2 \frac{\kappa_n^m}{2}}. \end{aligned} \tag{21}$$

Suppose that Γ_∞ and $\Gamma_{-\infty}$ are either 1 *or* $1 + \frac{\varepsilon^4}{\delta^2}$. *Then, it follows that*

1. $|\gamma_{n+1}^{m+1} - \gamma_n^{m+1}| = \varepsilon$. Namely, γ_n^{m+1} is an isoperimetric deformation of γ_n^m.
2. u_n^{m+1} gives the complex discrete curvature of γ_n^{m+1} (Fig. 1).

Remark 1 1. The deformation (21) is *not* an equidistant deformation in contrast with the deformation described by dmKdV [11]. In fact, one can show that

$$\left|\frac{\gamma_n^{m+1} - \gamma_n^m}{\delta}\right|^2 = \frac{4}{\varepsilon^2}\left(-1 + \frac{\Gamma_n^m}{\cos^2 \frac{\kappa_n^m}{2}}\right). \tag{22}$$

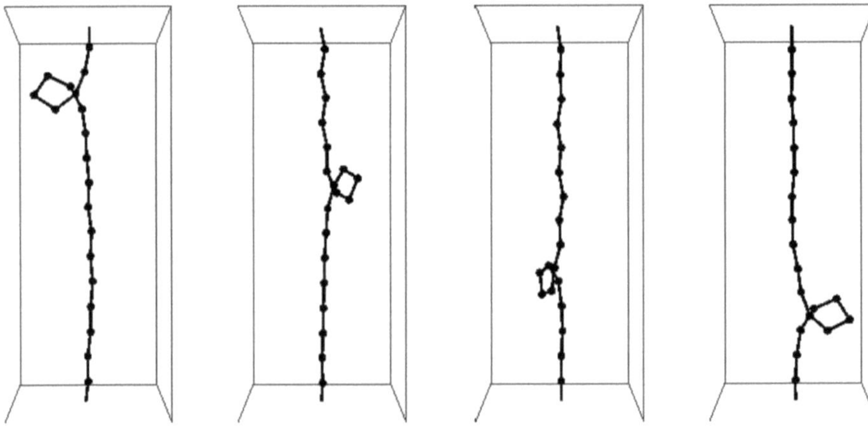

Fig. 1 Numerical simulation of dNLS flow

Equation (22) also implies that the solution of dNLS (15) should satisfy the condition $\Gamma_n^m \geq \cos^2 \frac{\kappa_n^m}{2}$ in order to be consistent with the curve deformation. Note that this property does not contradict with Hoffmann's formulation where the deformation is constructed as composition of two isoperimetric and equidistant deformations.

2. Continuous limit with respect to time can be simply taken as $t = m\delta$ and $\delta \to 0$. Then dNLS (15) and corresponding deformation equation (20) and (21) yields the *semi-discrete NLS equation* or the *Ablowitz–Ladik equation* [1, 2]

$$\sqrt{-1}\frac{du_n}{dt} + \frac{u_{n+1} - 2u_n + u_{n-1}}{\varepsilon^2} + (u_{n+1} + u_{n-1})|u_n|^2 = 0, \tag{23}$$

and the deformation equation of discrete space curves [3, 7, 15]

$$\frac{d}{dt}\gamma_n = \frac{2}{\varepsilon} \tan \frac{\kappa_n}{2} B_n. \tag{24}$$

The dNLS flow (20) and (21) implies the deformation of Frenet frame as

$$\Phi_n^{m+1} = \Phi_n^m M_n^m,$$

$$M_n^m = \frac{1}{\Gamma_{n+1}^m}\begin{bmatrix} |\alpha_n^m|^2 - |\beta_n^m|^2 & 2\Re\left(\alpha_n^m \beta_n^{m*}\right) & -2\Im\left(\alpha_n^m \beta_n^{m*}\right) \\ -2\Re\left(\alpha_n^m \beta_n^m\right) & \Re\left(\alpha_n^{m2} - \beta_n^{m2}\right) & -\Im\left(\alpha_n^{m2} + \beta_n^{m2}\right) \\ -2\Im\left(\alpha_n^m \beta_n^m\right) & \Im\left(\alpha_n^{m2} - \beta_n^{m2}\right) & \Re\left(\alpha_n^{m2} + \beta_n^{m2}\right) \end{bmatrix} \in \mathrm{SO}(3), \tag{25}$$

where $\alpha_n^m, \beta_n^m \in \mathbb{C}$ are given by

$$\alpha_n^m = \sqrt{-1}\frac{\delta}{\varepsilon^2}\left[\left(1 - \sqrt{-1}\frac{\varepsilon^2}{\delta}\right) - \left(1 + \varepsilon^2 u_{n+1}^m u_n^{m+1*}\right)\Gamma_{n+1}^m\right] e^{\frac{\sqrt{-1}}{2}(\Lambda_n^{m+1} - \Lambda_n^m)},$$

$$\beta_n^m = \sqrt{-1}\frac{\delta}{\varepsilon}\left(u_n^{m+1} - u_{n+1}^m\right)\Gamma_{n+1}^m e^{-\frac{\sqrt{-1}}{2}(\Lambda_n^m + \Lambda_n^{m+1})}, \tag{26}$$

respectively. Here, $*$ means the complex conjugate. The Frenet–Serret formula (19) and the deformation equation (25) can be transformed to the SU(2) version by the standard correspondence of SO(3) and SU(2) as

$$\phi_{n+1}^m = \phi_n^m L_n^m, \quad L_n^m = \begin{bmatrix} \cos\frac{\kappa_{n+1}^m}{2} e^{-\frac{\sqrt{-1}}{2}\nu_{n+1}^m} & -\sin\frac{\kappa_{n+1}^m}{2} e^{-\frac{\sqrt{-1}}{2}\nu_{n+1}^m} \\ \sin\frac{\kappa_{n+1}^m}{2} e^{\frac{\sqrt{-1}}{2}\nu_{n+1}^m} & \cos\frac{\kappa_{n+1}^m}{2} e^{\frac{\sqrt{-1}}{2}\nu_{n+1}^m} \end{bmatrix},$$

$$\phi_n^{m+1} = \phi_n^m M_n^m, \quad M_n^m = \frac{1}{\sqrt{\Gamma_{n+1}^m}}\begin{bmatrix} \alpha_n^m & \beta_n^m \\ -\beta_n^{m*} & \alpha_n^{m*} \end{bmatrix}, \tag{27}$$

which is known as the *Lax pair* of dNLS [1, 2]. In fact, one can verify that the compatibility condition $L_n^m M_{n+1}^m = M_n^m L_n^{m+1}$ yields dNLS (15).

Outline of the Proof of Theorem 1

The first statement may be verified directly in principle, by computing $\gamma_{n+1}^{m+1} - \gamma_n^{m+1}$ and its length from (20) and (21) and the discrete Frenet–Serret formula (19) under the assumption that u_n^{m+1} is determined by dNLS (15). However, this computation is hopelessly complicated to carry out. To make it feasible, we change the Frenet frame to a different frame used in [6, 13], which we call the *complex parallel frame* in this paper. Let $F_n^m = [T_n^m, U_n^m, U_n^{m*}] \in \mathrm{U}(3)$ be the complex parallel frame defined by

$$U_n^m = \frac{e^{\sqrt{-1}\Lambda_n^m}}{\sqrt{2}}\left(N_n^m + \sqrt{-1}B_n^m\right). \tag{28}$$

Note that it is related to the discrete Frenet frame Φ_n^m as

$$F_n^m = \Phi_n^m \begin{bmatrix} 1 & 0 & 0 \\ 0 & 1 & 1 \\ 0 & \sqrt{-1} & -\sqrt{-1} \end{bmatrix} \begin{bmatrix} 1 & 0 & 0 \\ 0 & \frac{e^{\sqrt{-1}\Lambda_n^m}}{\sqrt{2}} & 0 \\ 0 & 0 & \frac{e^{-\sqrt{-1}\Lambda_n^m}}{\sqrt{2}} \end{bmatrix}. \tag{29}$$

Then the complex curvature u_n^m naturally arises in this framework; the discrete Frenet–Serret formula (19) and the deformation of the discrete curve are rewritten in terms of u_n^m as

$$F_{n+1}^m = F_n^m X_n^m,\ X_n^m = \frac{1}{1+\varepsilon^2\left|u_{n+1}^m\right|^2}\begin{bmatrix} 1-\varepsilon^2\left|u_{n+1}^m\right|^2 & -\sqrt{2}\varepsilon u_{n+1}^m & -\sqrt{2}\varepsilon {u_{n+1}^m}^* \\ \sqrt{2}\varepsilon {u_{n+1}^m}^* & 1 & -\varepsilon^2({u_{n+1}^m}^*)^2 \\ \sqrt{2}\varepsilon u_{n+1}^m & -\varepsilon^2(u_{n+1}^m)^2 & 1 \end{bmatrix}, \tag{30}$$

and

$$\gamma_n^{m+1} = \gamma_n^m + \frac{2\delta^2}{\varepsilon^3} F_n^m \begin{bmatrix} -1+\left(1+\varepsilon^2\left|u_n^m\right|^2\right)\Gamma_n^m \\ \frac{\varepsilon}{\sqrt{2}}\left\{\left(1-\sqrt{-1}\frac{\varepsilon^2}{\delta}\right)u_n^{m*} - u_{n-1}^{m+1*}\left(1+\varepsilon^2\left|u_n^m\right|^2\right)\Gamma_n^m\right\} \\ \frac{\varepsilon}{\sqrt{2}}\left\{\left(1+\sqrt{-1}\frac{\varepsilon^2}{\delta}\right)u_n^m - u_{n-1}^{m+1}\left(1+\varepsilon^2\left|u_n^m\right|^2\right)\Gamma_n^m\right\} \end{bmatrix}, \tag{31}$$

respectively. The following lemma plays a crucial role in the proof:

Lemma 1 *Let $\gamma_n^m \in \mathbb{R}^3$ be the family of discrete space curves given in Theorem 1. Then it follows that*

$$|\alpha_n^m|^2 + |\beta_n^m|^2 = \Gamma_{n+1}^m. \tag{32}$$

By using (32), we have after long but straightforward calculations

$$T_n^{m+1} = \frac{\gamma_{n+1}^{m+1} - \gamma_n^{m+1}}{\varepsilon} = F_n^m \frac{1}{\Gamma_{n+1}^m}\begin{bmatrix} \left|\alpha_n^m\right|^2 - \left|\beta_n^m\right|^2 \\ -\sqrt{2}\alpha_n^{m*}\beta_n^{m*}e^{-\sqrt{-1}\Lambda_n^m} \\ -\sqrt{2}\alpha_n^m\beta_n^m e^{\sqrt{-1}\Lambda_n^m} \end{bmatrix}, \tag{33}$$

from which we obtain

$$\left|\frac{\gamma_{n+1}^{m+1} - \gamma_n^{m+1}}{\varepsilon}\right|^2 = \frac{\left(\left|\alpha_n^m\right|^2 - \left|\beta_n^m\right|^2\right)^2 + 4\left|\alpha_n^m\right|^2\left|\beta_n^m\right|^2}{{\Gamma_{n+1}^m}^2} = \left(\frac{\left|\alpha_n^m\right|^2 + \left|\beta_n^m\right|^2}{\Gamma_{n+1}^m}\right)^2 = 1.$$

This proves the first statement. The second statement is proved as follows. Starting from T_n^{m+1} (33), we have B_n^{m+1} and N_n^{m+1} in terms of F_n^m by using (11). Then we obtain an expression of $\Phi_n^{m+1} = [T_n^{m+1}, N_n^{m+1}, B_n^{m+1}]$ in terms of F_n^m, which can be rewritten as $F_n^{m+1} = F_n^m Y_n^m$ with a certain matrix $Y_n^m \in \mathrm{U}(3)$ by using (29). This can be also transformed to the deformation equation of discrete Frenet frame of the form $\Phi_n^{m+1} = \Phi_n^m M_n^m$ with M_n^m given in (25). Finally one can check that Φ_n^{m+1} satisfies the discrete Frenet–Serret formula (19) for κ_n^{m+1} and ν_n^{m+1} determined from the complex curvature u_n^{m+1}. This completes the proof of Theorem 1. □

3 Explicit Formulas

The formulation of NLS and dNLS flows in terms of the Frenet frame enables us to apply the theory of integrable systems. As an example, we here present explicit formulas of the NLS and dNLS flows in terms of the τ functions. For the case of plane curves, see [10]. These formulas are established based on the bilinear formalism in the theory of integrable systems by applying suitable reductions and imposing complex structure to τ functions of the 2-component KP hierarchy, but here we only show the results, leaving full derivations to the forthcoming publications.

For any $N \in \mathbb{N}$, we first introduce the following three determinants, a $2N \times 2N$ determinant τ, two $(2N+1) \times (2N+1)$ determinants σ and ρ as

$$\tau = \left|\begin{array}{ccc|ccc} m_{11}^{(1)} & \cdots & m_{1N}^{(1)} & 1 & & \emptyset \\ \vdots & \cdots & \vdots & & \ddots & \\ m_{N1}^{(1)} & \cdots & m_{NN}^{(1)} & \emptyset & & 1 \\ \hline -1 & & \emptyset & m_{11}^{(2)} & \cdots & m_{1N}^{(2)} \\ & \ddots & & \vdots & \cdots & \vdots \\ \emptyset & & -1 & m_{N1}^{(2)} & \cdots & m_{NN}^{(2)} \end{array}\right|, \tag{34}$$

$$\sigma = \left|\begin{array}{ccc|ccc|c} m_{11}^{(1)} & \cdots & m_{1N}^{(1)} & 1 & & \emptyset & \varphi_1^{(1)} \\ \vdots & \cdots & \vdots & & \ddots & & \vdots \\ m_{N1}^{(1)} & \cdots & m_{NN}^{(1)} & \emptyset & & 1 & \varphi_N^{(1)} \\ \hline -1 & & \emptyset & m_{11}^{(2)} & \cdots & m_{1N}^{(2)} & 0 \\ & \ddots & & \vdots & \cdots & \vdots & \vdots \\ \emptyset & & -1 & m_{N1}^{(2)} & \cdots & m_{NN}^{(2)} & 0 \\ \hline 0 & \cdots & 0 & \varphi_1^{(2)} & \cdots & \varphi_N^{(2)} & 0 \end{array}\right|, \tag{35}$$

$$\rho = \left|\begin{array}{ccc|ccc|c} m_{11}^{(1)} & \cdots & m_{1N}^{(1)} & 1 & & \emptyset & 0 \\ \vdots & \cdots & \vdots & & \ddots & & \vdots \\ m_{N1}^{(1)} & \cdots & m_{NN}^{(1)} & \emptyset & & 1 & 0 \\ \hline -1 & & \emptyset & m_{11}^{(2)} & \cdots & m_{1N}^{(2)} & \psi_1^{(1)} \\ & \ddots & & \vdots & \cdots & \vdots & \vdots \\ \emptyset & & -1 & m_{N1}^{(2)} & \cdots & m_{NN}^{(2)} & \psi_N^{(1)} \\ \hline \psi_1^{(2)} & \cdots & \psi_N^{(2)} & 0 & \cdots & 0 & 0 \end{array}\right|, \tag{36}$$

where Ø is the empty block. Then the formulas for NLS flow on smooth curves and dNLS flow on discrete curves are obtained by choosing the entries of determinant as follows:

NLS Flow on Smooth Curves

We choose the entries of determinants as

$$\begin{aligned} &\varphi_i^{(1)} = p_i^n e^{\xi_i}, \quad \varphi_i^{(2)} = -1, \\ &\psi_i^{(1)} = 1, \quad \psi_i^{(2)} = -\left(-\frac{1}{p_i^*}\right)^n e^{\xi_i^*}, \\ &m_{ij}^{(1)} = -\frac{\varphi_i^{(1)}\psi_j^{(2)}}{p_i + p_j^*}, \quad m_{ij}^{(2)} = \frac{1}{p_i^* + p_j}, \\ &\xi_i = p_i x - \sqrt{-1} p_i^2 t + \frac{1}{p_i} z + \xi_i^{(0)}, \quad p_i, \ \xi_i^{(0)} \in \mathbb{C}, \end{aligned} \tag{37}$$

so that we write $\tau = \tau_n(x, t; z)$, $\sigma = \sigma_n(x, t; z)$, $\rho = \rho_n(x, t; z)$. Here, n and z are regarded as auxiliary variables. Putting

$$F = \tau_0, \quad G = -\rho_0, \quad h = -\sigma_{-2}, \tag{38}$$

we have:

Theorem 2 (Explicit formula for NLS flow)[1]

1. *Let $u = u(x, t) \in \mathbb{C}$ be*

$$u = 2\frac{G}{F}. \tag{39}$$

   Then u satisfies NLS (9).
2. *Let $\gamma = \gamma(x, t) \in \mathbb{R}^3$ be*

$$\gamma = \begin{bmatrix} \dfrac{h + h^*}{F} \\ \dfrac{1}{\sqrt{-1}} \dfrac{h - h^*}{F} \\ 2\dfrac{\partial}{\partial z}(\log F) - x \end{bmatrix}. \tag{40}$$

   Then γ satisfies the Frenet–Serret formula (3) and the deformation equation (4) (Fig. 2).

[1]The N-soliton solution for the tangent vector has been constructed by using the bilinear formalism in [5].

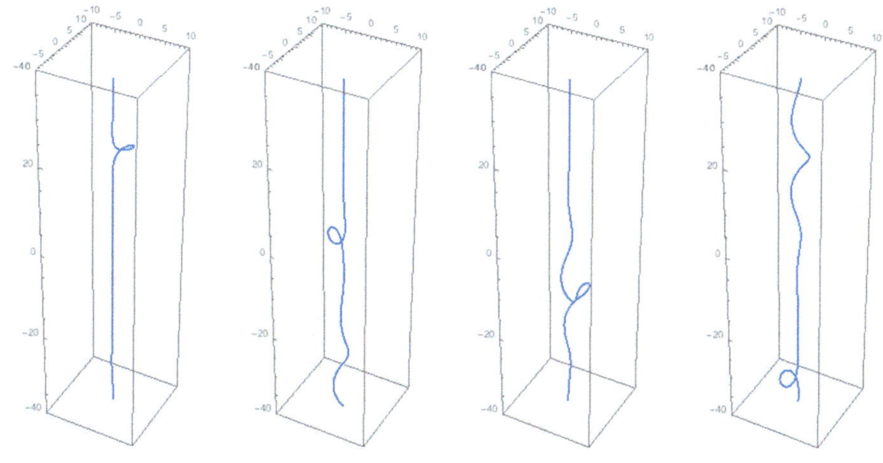

Fig. 2 Interaction of loops of smooth curve by NLS flow obtained from Theorem 2

dNLS flow on discrete curves

We choose the entries of determinants as

$$
\begin{aligned}
&\varphi_i^{(1)} = p_i^{-n} e^{\zeta_i} (1 - a p_i)^{-m} (1 - c p_i)^{-r}, \ \varphi_i^{(2)} = -\left(1 - \tfrac{a}{p_i}\right)^m \left(1 - \tfrac{1}{c p_i}\right)^s, \\
&\psi_i^{(1)} = (1 - a p_i^*)^{-m} \left(1 - \tfrac{p_i^*}{c}\right)^{-s}, \ \psi_i^{(2)} = -(p_i^*)^{-n} e^{\zeta_i^*} \left(1 - \tfrac{a}{p_i^*}\right)^m \left(1 - \tfrac{c}{p_i^*}\right)^r, \\
&m_{ij}^{(1)} = -\frac{\varphi_i^{(1)} \psi_j^{(2)}}{p_i - \frac{1}{p_j^*}}, \quad m_{ij}^{(2)} = -\frac{\psi_i^{(1)} \varphi_j^{(2)}}{p_i^* - \frac{1}{p_j}}, \quad e^{\zeta_i} = e^{\frac{1}{2} \frac{1 + c p_i}{1 - c p_i} z}, \\
&\qquad a \in \mathbb{R}, \quad p_i, \ c \in \mathbb{C}, \quad |c| = 1,
\end{aligned}
\tag{41}
$$

so that we write $\tau = \tau_n^m(r, s; z)$, $\sigma = \sigma_n^m(r, s; z)$, $\rho = \rho_n^m(r, s; z)$ with r, s and z being auxiliary variables. Putting

$$
F_n^m = \tau_n^m(0, 0; z), \quad G_n^m = \rho_n^m(0, 0; z), \quad h_n^m = c^{-n} \sigma_n^m(1, -1; z), \tag{42}
$$

$$
a = \left(1 + \frac{\varepsilon^4}{\delta^2}\right)^{\frac{1}{2}}, \quad c = \frac{1 - \sqrt{-1}\frac{\varepsilon^2}{\delta}}{a}, \tag{43}
$$

we have:

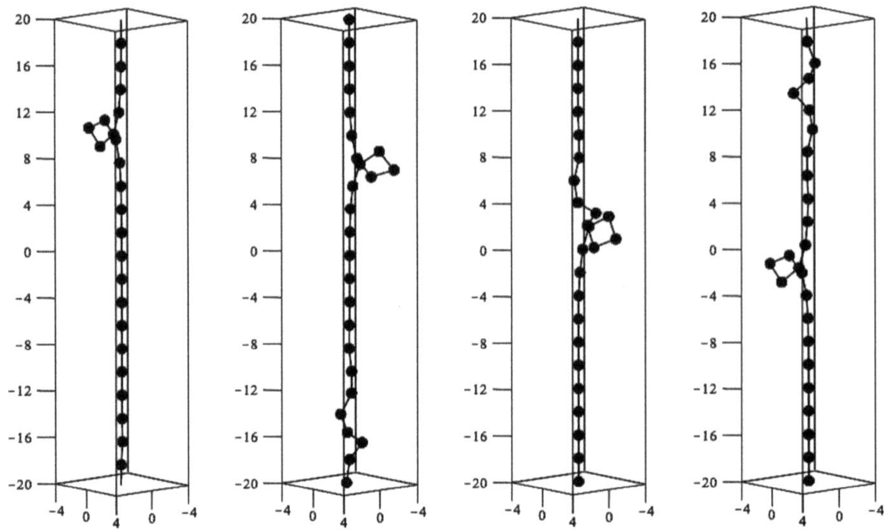

Fig. 3 Interaction of loops of discrete curve by dNLS flow obtained from Theorem 3

Theorem 3 (Explicit formula for dNLS flow)

1. *Let* $u_n^m \in \mathbb{C}$ *be*

$$u_n^m = \frac{(-1)^m c^{-n-2m}}{\varepsilon} \frac{G_n^m}{F_n^m}, \quad \Gamma_n^m = \frac{2a}{c+\frac{1}{c}} \frac{F_{n-1}^{m+1} F_n^m}{F_n^{m+1} F_{n-1}^m}. \tag{44}$$

Then u_n^m *satisfies dNLS (15).*

2. *Let* $\gamma_n^m \in \mathbb{R}^3$ *be*

$$\gamma_n^m = \varepsilon \begin{bmatrix} (-1)^m \dfrac{h_n^m + h_n^{m*}}{F_n^m} \\ \dfrac{(-1)^m}{\sqrt{-1}} \dfrac{h_n^m - h_n^{m*}}{F_n^m} \\ 2\dfrac{\partial}{\partial z}(\log F_n^m) - n - 2m \end{bmatrix}. \tag{45}$$

Then γ_n^m *satisfies the Frenet–Serret formula (19) and the deformation equation (20) and (21) (Fig. 3).*

References

1. M.J. Ablowitz, J.F. Ladik, A nonlinear difference scheme and inverse scattering. Stud. Appl. Math. **55**, 213–229 (1976)
2. M.J. Ablowitz, J.F. Ladik, On the solution of a class of nonlinear partial difference equations. Stud. Appl. Math. **57**, 1–12 (1977)
3. A. Doliwa, P.M. Santini, Integrable dynamics of a discrete curve and the Ablowitz-Ladik hierarchy. J. Math. Phys. **36**, 1259–1273 (1995)
4. A. Doliwa, P.M. Santini, The integrable dynamics of a discrete curve, in *Symmetries and Integrability of Difference Equations*, vol. 9, CRM Proceedings & Lecture Notes, ed. by D. Levi, L. Vinet, P. Winternitz (American Mathematical Society, Providence, 1996), pp. 91–102
5. Y. Fukumoto, T. Miyazaki, N-Solitons on a curved vortex filament. J. Phys. Soc. Jpn. **55**, 4152–4155 (1986)
6. H. Hasimoto, Soliton on a vortex filament. J. Fluid Mech. **51**, 477–485 (1972)
7. M. Hisakado, M. Wadati, Moving discrete curve and geometric phase. Phys. Lett. A **214**, 252–258 (1996)
8. T. Hoffmann, On the equivalence of the discrete nonlinear Schrödinger equation and the discrete isotropic Heisenberg magnet. Phys. Lett. A **265**, 62–67 (2000)
9. T. Hoffmann, Discrete Hashimoto surfaces and a doubly discrete smoke-ring flow, in *Discrete Differential Geometry*, vol. 38, Oberwolfach Seminars, ed. by A.I. Bobenko, P. Schröder, J.M. Sullivan, G.M. Ziegler (Birkhäuser, Basel, 2008), pp. 95–115
10. J. Inoguchi, K. Kajiwara, N. Matsuura, Y. Ohta, Motion and Bäcklund transformations of discrete plane curves. Kyushu J. Math. **66**, 303–324 (2012)
11. J. Inoguchi, K. Kajiwara, N. Matsuura, Y. Ohta, Discrete mKdV and discrete sine-Gordon flows on discrete space curves. J. Phys. A: Math. Theor. **47**, 235202 (2014)
12. J. Inoguchi, K. Kajiwara, N. Matsuura, Y. Ohta, Discrete models of isoperimetric deformation of plane curves, in *Mathematical Progress in Expressive Image Synthesis*, vol. 4, Mathematics for Industry, ed. by K. Anjyo (Springer, Tokyo, 2014), pp. 111–122
13. G.L. Lamb, Solitons on moving space curves. J. Math. Phys. **18**, 1654–1659 (1977)
14. N. Matsuura, Discrete KdV and discrete modified KdV equations arising from motions of planar discrete curves. Int. Math. Res. Not. **2012**, 1681–1698 (2012)
15. K. Nakayama, Elementary vortex filament model of the discrete nonlinear Schrödinger equation. J. Phys. Soc. Jpn. **76**, 074003 (2007)
16. K. Nakayama, H. Segur, M. Wadati, Integrability and the motions of curves. Phys. Rev. Lett. **69**, 2603–2606 (1992)
17. U. Pinkall, B. Springborn, S. Weißmann, A new doubly discrete analogue of smoke ring flow and the real time simulation of fluid flow. J. Phys. A: Math. Theor. **40**, 12563–12576 (2007)
18. S. Tsujimoto, Discretization of integrable systems, in *Applied Integrable Systems*, ed. by Y. Nakamura (Shokabo, Tokyo, 2000), pp. 1–52 (In Japanese)
19. S. Weißmann, U. Pinkall, Real-time interactive simulation of smoke using discrete integrable vortex filaments, in *VRIPhys: Workshop on Virtual Reality Interaction and Physical Simulations*, ed. by H. Prautzsch, et al. (The Eurographics Association, Aire-la-Ville, 2009), pp. 1–10

Index

Y. Dobashi and H. Ochiai (eds.), *Mathematical Progress in Expressive Image Synthesis III*, Mathematics for Industry 24,
DOI 10.1007/978-981-10-1076-7

MIX
Papier aus verantwortungsvollen Quellen
Paper from responsible sources
FSC® C105338

If you have any concerns about our products,
you can contact us on
ProductSafety@springernature.com

In case Publisher is established outside the EU,
the EU authorized representative is:
Springer Nature Customer Service Center GmbH
Europaplatz 3, 69115 Heidelberg, Germany

Printed by Libri Plureos GmbH
in Hamburg, Germany